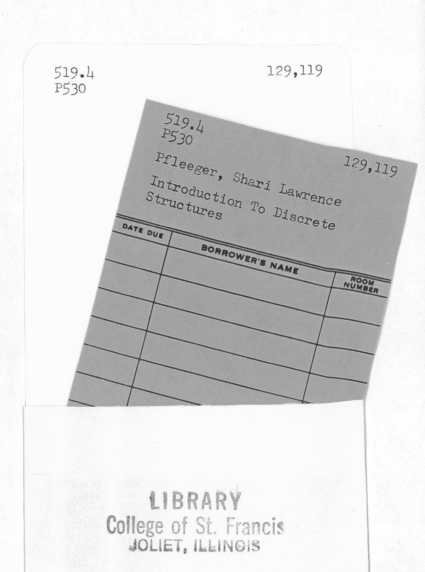

INTRODUCTION TO
DISCRETE STRUCTURES

INTRODUCTION TO DISCRETE STRUCTURES

SHARI LAWRENCE PFLEEGER
Systems/Software Inc.

DAVID W. STRAIGHT
Department of Computer Science
University of Tennessee

JOHN WILEY & SONS
New York · Chichester · Brisbane · Toronto · Singapore

Cover art: Paul Klee, "Old Klang," 1925, Giraudon/Art Resource

Library of Congress Cataloging in Publication Data:

Pfleeger, Shari Lawrence.
    Introduction to discrete structures.

    Includes index.
    1. Electronic data processing—Mathematics.
I. Straight, David W., 1942–   . II. Title.

QA76.9.M35P45 1985   519.4   84-21005
ISBN 0-471-80075-9

Printed in the United States of America

10 9 8 7 6 5 4 3 2 1

PREFACE

This book presents the core material for a one-semester or one-quarter course in discrete structures for an undergraduate interested in computer science. Such a course is part of the standard ACM curriculum, and the majority of students taking the course will be computer science majors. In the past, a computer science curriculum comprised courses on hardware, software, and theory. Because some of the theory has mathematical origins, the discrete structures courses were often taught by members of the mathematics faculty. Indeed, many computer science departments had as their origin a special interest group in a mathematics department. Today, the situation has changed, and a computer science major need not be immersed in mathematics to study computer science. The prerequisites for a discrete structures course vary from requiring no background other than high school algebra and geometry to a strong mathematics background including several courses in calculus.

We strongly believe that one of the purposes of a discrete structures course is to provide the student with a certain degree of mathematical maturity and an ability to work with abstract concepts. It is possible that the student has already acquired this maturity in previous mathematics courses, but it is not at all necessary. Both of us have strong mathematical backgrounds and have a great love of the theoretical aspects of computer science. However, along with developing in the students a similar love, we are committed to showing the student how the theory ties together concepts learned in studying hardware and software. To this end, this book requires no mathematical background beyond high school algebra and geometry. Furthermore, it combines theory with applications. Wherever possible, the student is shown how discrete structures form an underlying framework for much that is considered computer science. In this way, the study of discrete structures is shown to be relevant to the day-to-day concerns of programmers, analysts, and computer scientists. At the same time, the student learns to work in a rigorous, formally defined theoretical system.

Students often wish to learn skills that are "portable"; that is, they appreciate best those concepts that are applicable to many of the situations with which they are familiar. It is easier to keep students' attention when they can see how the concepts are relevant than when the theory stands by itself. Because of this, we begin the book with a Chapter introducing students to the concepts to be studied while giving them a taste of the kinds of problems to which these concepts can be applied. The book does not promise to unveil all of the answers to the problems described in Chapter 0, but it points students in the right direction and to the right tools.

Chapter 1 provides some of the basic building blocks of what is to come: sets, truth and validity, and proof techniques. Chapter 2 moves on to introduce functions and relations. Using the material from the first two chapters, Chapter 3 defines relations on sets and shows how Boolean algebras can be formed. Chapters 4 and 5 move from there to lattices, and this knowledge is used to explain how state diagrams can be generated and reduced using Karnaugh maps, the Quine-McCluskey method, and other techniques. In Chapter 6, the student learns about permutations, combinations, and counting techniques; in particular, partitioning and counting are shown to be useful techniques in proving theorems.

Graph theory is explored in Chapter 7, and the student is taught to visualize computer science concepts using directed and undirected graphs. Network traversal, map coloring, maximal flow, and other problems are described in graph-theoretic terms, and their solutions using a computer are introduced. Chapter 8 moves on to discuss concepts that involve language and shows how languages and grammars are related to computing. Finally, Chapter 9 addresses problems of computability. The exercises at the end of each chapter are designed both to test the students' mastery of the concepts introduced and to challenge the students with programming applications of those concepts. The completion of the course of study outlined should provide students not only with an understanding of the theory but also an appreciation of how intricately intertwined are discrete structures in the fiber of computer science.

Shari Lawrence Pfleeger
David W. Straight

Knoxville, Tennessee
June 1984

ACKNOWLEDGMENTS

Many thanks are due to friends and colleagues who offered support and advice during the writing of this book. First and foremost, we are grateful to Charles Pfleeger for his encouragement, professional expertise, moral support, and friendship. The insightful suggestions of Charles Hughes were greatly appreciated, too. We owe a great deal to Florence Rogart, Raymond Ayoub and Norman Martin, the mentors who revealed to us the beauty and elegance of mathematics.

This book would never have been completed without the diligence and patience of Francine Stonehouse (undeterred by a broken leg and a move to Chattanooga) and Suzie Wittenbarger. They sacrificed innumerable evenings and weekends to type our manuscript. They were ably assisted by Ethel Wittenberg and Sally Collins. Mary Bigger provided the answers to the exercises for the teachers' supplement, and Nancy Friese was kind enough to give us permission to use some of her artwork.

Finally, we thank the University of Tennessee Computing Center and Systems/Software, Inc. for the use of their computing facilities. Systems/Software also arranged for release time to prepare the manuscript.

S.L.P.
D.W.S.

Credits for Chapter Opening Art

CHAPTER 0: Fernand Léger, "Les Constructeurs." Courtesy F. Léger Estate. Reprinted by permission.

CHAPTER 1: Georges Seurat, "Channel at Gravelines," 1890. Collection, The Museum of Modern Art, New York, gift of William A. M. Burden.

CHAPTER 2: Pavel Tchelitchew, "Head VI." Collection, The Museum of Modern Art, New York, gift of Edgar Kaufmann, Jr.

CHAPTER 3: C. D. Friedrich, "The Church Yard." Kunsthalle, Bremen. Reprinted by permission.

CHAPTER 4: Piet Mondrian, "Broadway Boogie Woogie." Collection, The Museum of Modern Art, New York, given anonymously.

CHAPTER 5: M. C. Escher, "Convex and Concave." Collection Haags Gemeentemuseum The Hague.

CHAPTER 6: Charles Demuth, "Business." The Alfred Stieglitz Collection, courtesy The Art Institute of Chicago.

CHAPTER 7: Nancy Friese, "To The Western Hills." Reprinted by permission of the artist.

CHAPTER 8: Pieter Breughel, "Tower of Babel." Kunsthistorisches Museum, Vienna.

CHAPTER 9: Charles Sheeler, "Rolling Power." Smith College Museum of Art, Northampton, Massachusetts.

CONTENTS

CHAPTER 8 INTRODUCTION TO FORMAL LANGUAGES   285

CHAPTER 9 COMPUTABILITY   321

CHAPTER 0

WHY YOU SHOULD READ THIS BOOK

As a student of computer science, it is likely that you understand why courses about such topics as compiler construction, data base management, and programming languages are important parts of your education about computing. Just where discrete structures fit in is probably not so clear. To address this problem, this book will have two equally important emphases: what discrete structures are, and how they are embedded in the science of computing. As you will see, discrete structures play an essential role in almost all areas of computing. We hope that you will view your course in discrete structures as part of the theoretical basis of computer science, rather than as peripheral and not very relevant compared with some other parts of your curriculum.

WHAT ARE DISCRETE STRUCTURES?

Let us look more closely at exactly what is meant by "discrete structures." To understand the term "discrete," consider two kinds of printing devices. The first kind can draw a line from one point to another by putting a pen down at one point and keeping it down as the pen slides across the paper to the second point. The result may look something like Figure 0.1. The second printing device can only print dots. Like your typewriter or a dot matrix printer, this second printing device is an impact device. It hits the paper with a printing head and then retracts. To draw a line from one point to the next, the second printer must approximate the line with a series of dots. The result would look something like Figure 0.2. We say that the line

Figure 0.1   Figure 0.2

drawn by the first printer is a *continuous* line, and the line drawn by the second is *discrete*. This is because there are no gaps in the first line but many in the second (even if the impact printer prints dots so fine that they are imperceptible to a human eye).

In the same way, there are two kinds of computer, *digital* and *analog*. In a digital computer, values are represented as sequences of bits, where a bit is a variable that is either zero or 1. In a computer that uses 16 bits to represent numbers, we may have two numbers represented as

0110001101111000

and

0110001101111001

Note that these two numbers differ only by a single bit in the rightmost position. There is a jump or discrete step between the two values. Because of this manner of representing numbers, it is impossible to represent all values on a digital machine. After having chosen a representation scheme (such as the integers in base 2), we can represent at most 65,536 different values using 16 bits. Other values, such as 1/3 or the square root of 2, cannot be represented but can be approximated by this 16−bit base−2 representation. Thus, a digital machine cannot represent every number exactly.

Analog computers, on the other hand, represent values by using something which can take on a continuous range of values. For instance, a typical analog computer will represent a number by using voltages. Thus, any rational number (such as 1/3) or irrational number (such as pi and the square root of 2) can easily be represented on an analog computer. However, analog computers present severe problems with precision. The precision is limited by the capabilities of the measuring devices. One might have a voltmeter that can measure voltage accurately to three decimal places, but a voltmeter that measures up to 20 decimal places may be impossible to construct using today's technology. In spite of the precision problem, analog computers can be very useful when working with continuous functions or differential equations.

Digital computers are far more common than analog ones. Although digital computers do not have the full range of values offered by analogs, the precision of a digital computer is far better at the selected points it can represent. Albeit with some additional effort, a value can be represented to any finite number of decimal places. Thus, whereas an analog computer is a *continuous* machine, the values on a digital computer are *discrete*.

Discrete structures, then, deal with the ways in which various kinds of data can be *represented* and *manipulated* in a discrete or digital manner. Not only are the representations considered discrete structures, but so are the relationships among these representations. As such, the study of discrete structures is the study of the theoretical and foundational material of computer science.

HOW CAN WE USE DISCRETE STRUCTURES?

Many of you are studying computer science in order to become professional computer scientists. Hence, you may have a particular interest in the application areas, such as data base management or compiler writing. You will find that a great many of the applications you will encounter have their roots in discrete structures. Not only will a knowledge of discrete structures enable you to better understand current applications of computing, but it will also enhance your ability to deal with new applications.

Because we feel so strongly about the usefulness of this field, this chapter is an introduction to the kinds of problems to which discrete structures can be applied. Rather than bombard you with technical explanations from the beginning, we would like to give you a feeling for the way in which discrete structures crop up in daily situations (and especially on the job).

INFINITE LOOP PROBLEM

Suppose that you are employed as a programmer. Your supervisor gives you a problem to solve:

Write a program called CHECKIT that will have as its input a PASCAL program. CHECKIT will read the PASCAL program, check if there are infinite loops in it, and return a code to the user: 0 if there are no infinite loops, 1 if there is at least one infinite loop.

On the face of it, this sounds like a reasonable problem. However, you hand these specifications back to your boss, saying "This can't be done." Why? Because using several basic principles of discrete structures, you can show that such a problem is not *solvable*. By applying what is known as *computability theory*, you can convince your boss that the problem can never be solved; you will save your company considerable time, effort, and expense.

SCHOOL BUS ROUTE PROBLEM

Suppose, instead, that you work for the board of education in your city. A school bus must pick up students at designated bus stops each morning. Clearly, it is in the best interest of everyone to have the driver take the shortest route, passing by each bus stop exactly once and minimizing the mileage driven. What is the best route for the school bus to take? Such a question really involves several questions, each of which can be expressed in terms of discrete structures:

1. For each pair of bus stops, what are all the different ways in which you can travel from one stop to the other?
2. Is there a path which visits each bus stop exactly once?
3. If there is such a path, is there more than one? And if so, how do you find the path of minimum mileage?
4. If such a path does not exist, how do you minimize your back—tracking?

An example of a school bus stop map is shown in Figure 0.3. The labels on the arcs indicate the mileage from one bus stop to another.

This kind of problem is called a *traveling sales representative problem*. In the way it is stated here, if the number of bus stops were small, you could probably sit down with a road atlas and work out the finite (and relatively small) number of solutions by hand. However, consider the kinds of constraints that can be placed on such a problem. The roads over which the bus can travel can vary in quality or usability (sometimes on a daily basis), depending on road construction projects, weather, and traffic accidents. This means that you may have to determine a new route whenever one of these constraints changes. Because you don't want the job of finding the

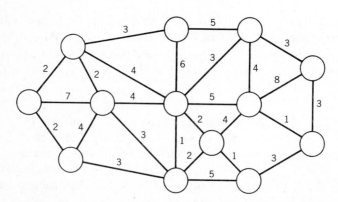

Figure 0.3

best route to become a lifetime occupation, you would certainly want to use a program to generate a solution; you could run the program whenever you are notified of a change in the description of the problem. Furthermore, there may be many different school bus routes. Instead of solving each problem independently, you would be better off to discover the common features of these problems and devise a general solution for all of the problems at once. Fortunately, discrete structures techniques help you to formulate this problem and tackle each of its component parts.

TRAFFIC FLOW PROBLEM

The traffic engineer for your city comes to you with a pressing problem. Over the last several years, the city university football team has developed a national reputation. Consequently, many out—of—towners attend the football games every Saturday. When the game is over, most of the traffic heads for the nearby interstate highway to leave town. All of the roads between the stadium and the highway entrance fill up and are jammed for hours. Although many of the city streets are made one—way streets temporarily to help alleviate the situation, there are still severe problems.

The diagram in Figure 0.4 represents the roads from the stadium to the highway entrance. Arrows indicate a one—way flow of traffic. The number assigned to each road represents the number of cars per minute that can be handled by that segment of road. (This is determined by such things as the width of the road, road condition, speed limits, etc.) How can you help the traffic engineer determine the maximal flow of traffic from the stadium to the highway entrance? In other words, how can you determine the maximum number of cars that can leave the stadium each minute and drive to

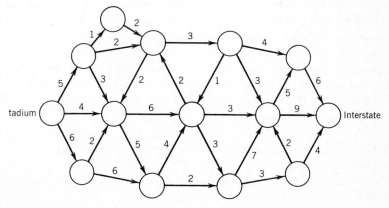

Figure 0.4

the interstate without causing a traffic jam? How can you locate the bottlenecks which might require the help of a traffic officer? If that can be determined, the highway engineer can then adjust the traffic lights to maintain a maximum flow. Such problems are known as *network flow problems.* We will see that discrete structures provide a framework for solving these problems, and algorithms are available for controlling the flow through the network.

MAXIMIZING PROFITS

Having impressed the traffic engineer with your ability to apply discrete structures, you are asked to do some consulting work for the traffic engineer's sister. She is president and chairperson of the board of the Irascible Widget Company. The company has three widget factories and two retail outlet stores. Raw materials are purchased from each of two sources, and you are given maximum and minimum production rates with each raw material at each factory. In addition, you are told the cost of each unit of raw material and the production costs at each factory. Finally, you are told the maximum and minimum sales at each retail outlet as well as the retail price per widget. Can you help Irascible Widget determine how much to buy, which factories to use, and how to distribute widgets to the retail outlets to maximize their profits? Indeed you can! Although more complex than the traffic flow problem, this is another example of a network flow.

COMPUTER COMMUNICATIONS NETWORK PROBLEM

For those of you with more of a "systems" than an "applications" interest, consider the following situation. You are working in the municipal computer center. The city wants to set up communications links among all of its computers so that different departments can trade information. You have been given a diagram showing the computers that need to be connected and the various possible connections that can be made. Associated with each possible connection is a number which indicates the relative cost of making that connection. Your task is to design a network that will join all of the computers as cheaply as possible. Thus, this problem can be decomposed into the following:

1. For each pair of computers, what are the ways in which these computers can communicate?
2. Is there a network that has exactly one connection between each pair of computers?

3. If such a network exists, is there more than one? And if so, how do you find the network of minimum cost?
4. If such a network does not exist, how can you find a network that will still minimize the number of connections?

This kind of problem can be viewed as in Figure 0.5 by applying *graph theory* techniques. The problem, then, is known as finding a *minimal cost spanning tree*. The circles are computers, and the numbers along the straight lines represent some measure of the cost of connecting the two computers. This problem is stated in simplified terms. In fact, it may be preferable, depending on costs, to route messages from computer A to computer C by sending them through computer B as an intermediary. (See Exercise 5.)

COMPILER CORRECTNESS PROBLEM

Once you have solved your computer network problem, you are given another assignment. You want to write a compiler, and you have carefully drawn up a list of requirements, that is, a list of all of the things that the compiler should be able to do. Then, you design the program KOMPYL, code it, and run it. How can you prove that KOMPYL does exactly what your requirements list specifies? Does the description of the language exactly match those programs that KOMPYL recognizes as correct and exclude those that KOMPYL rejects as incorrect?

Such a guarantee can prevent not only uncomfortable and embarrassing situations but also life—threatening ones. You may remember that several years ago, a NORAD computer in Colorado issued some false red alerts. Clearly, you should not be responsible for launching missiles or discharging troops by mistake because your compiler compiled the code incorrectly. If the military uses your compiler, they should be certain that it will do exactly

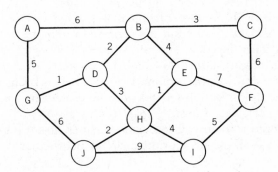

Figure 0.5

what you say it will do. If a bank uses your compiler, it wants to be sure that your compiler isn't causing interest rates to increase at random. If a university uses your compiler, it wants to know whether students will be able to change their grades or pay their bills illegally.

In general, this is typical of a kind of problem addressed by this book: How do you take specifications for something, build a program to meet those specifications, and ensure that the program works? It is in this sense that discrete structures provide a skeleton on which much of computer science is built.

LOGIC DESIGN PROBLEM

In designing a new computer, you may want to tackle the following problem: The plan for the new WARTHOG 4490 calls for 100,000 chips of a particular type. You are shown the logic design for the chip and told that there are 25 logic gates on each chip. Since each gate costs 1 cent, this means that each chip will cost 25 cents, thereby contributing a cost of $25,000 to the cost of the computer. Suppose your knowledge of discrete structures shows you a way to achieve the same results with only eight gates per chip. Then each chip will cost only eight cents, and you have thus reduced the cost of each computer by $17,000. Indeed, the use of logic and discrete structures can handle such problems, helping not only to slim down the circuits but perhaps even fatten your paycheck when your supervisor rewards you!

PRINTED CIRCUIT BOARD PROBLEM

Since you were so successful at the previous problem, the chief engineer asks you for help with another dilemma. A printed circuit board is to be used in the WARTHOG 4490, but the connections among components look like Figure 0.6. This design has a rather large number of crossovers. Can you redesign the figure above to eliminate as many of the crossovers as possible? Is there a way to determine if such a network can be drawn with no crossovers at all? In fact, a network with no crossovers can be represented in graph theory as a *planar graph*. In this book you will be writing computer programs to help you to decide whether or not a graph is planar. Figure 0.7 is an example of a printed circuit board that requires at least three planes.

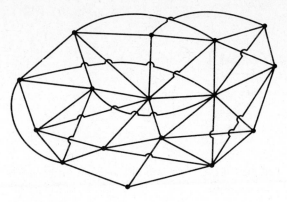

Figure 0.6

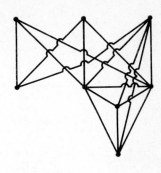

Figure 0.7

INSTANT INSANITY PROBLEM

If such a thing as "recreational discrete structures" exists, it may mean the application of discrete structures techniques to problems such as the "instant insanity" cubes with which you may be familiar. This game consists of four cubes, each having faces of red, blue, green and white arranged in various combinations, one color to each face. The object is to arrange the cubes in a row so that the same color does not appear twice in a row. If you have seen this puzzle and tried it out, you know that it can take hours to find a solution by simply rotating cubes, if you find one at all. However, a simple application of *graph theory* can solve the puzzle for you in just a few minutes.

MARATHON RUNNING ROUTE

A runner is to follow a prescribed course of 26 miles and 265 yards. Some of the more popular courses are in New York City, Boston, and San Francisco. The Gowanus Brewing Company would like to sponsor a marathon race in its home town of Gowanus City. To emphasize the fact that its beer is made with genuine Gowanus River water, the brewery wants the race to cross each of the many bridges that span the river. Using the map of Gowanus City shown in Figure 0.8, can you find a route that crosses each bridge exactly once? Is it possible to do that while starting and ending the race at the brewery? We will show that, using graph theory, the problem can be seen as one of traversing a graph under conditions similar to those of the famous *Königsberg bridge problem.* This problem is one which the well-known mathematician Euler solved for the townspeople of Königs-

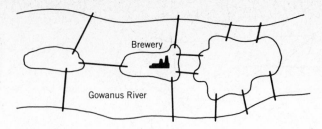

Figure 0.8

berg when they wanted to walk through their small German town each Sunday, crossing each bridge exactly once.

WHY STUDY DISCRETE STRUCTURES?

We hope that these descriptions give you some idea of the broad spectrum of applications of discrete structures. There is certainly no guarantee that the contents of this book will solve every problem you may meet in your everyday life or even in your life as computer scientist, but we believe that they will get you started. The first step is to *think* in terms of discrete structures, so that you can formulate the problems that confront you in simpler and more structured ways. Then, your knowledge of discrete structures will allow you to decide the following:

1. How can you tell what kinds of things you can write a program to solve?
2. If a problem is solvable, how do you find a way of using the computer (e.g., an algorithm) to solve it?
3. How do you convince others that your solution is indeed correct?

It is our intent to delve into these three questions plus a fourth: how do discrete structures relate to a wide number of other areas of computer science? To that end, nothing explained here will require knowledge of other than elementary algebra and geometry. Further, we will show how all material presented here has ties to other areas of computing.

EXERCISES

1. This chapter describes two very similar problems, that of routing school buses and that of designing a low—cost computer network. We have structured the problems in similar ways. Each problem can be expressed in terms of the cost of moving from one location to another. How do these problems differ from one another? If each type of problem could be depicted with the same network and the same costs, how would their solutions differ? Try this out with a small network.

2. The depiction of the marathon route problem is similar to that of the school bus routing problem. How does the solution to the marathon route determination problem differ from the solution to the school bus routing problem? As a hint, note that the marathon route must traverse each connection between points exactly once.

3. A cube has six sides, and each side can be numbered from 1 to 6 as on a set of dice. Suppose we align four dice so that we can see only one face of each cube, as in Figure 0.9. How many different configurations can we get? Now suppose that we are looking at the four dice from a slightly higher plane so that we can see two faces of the three leftmost cubes and three faces of the cube on the right end, as in Figure 0.10. How many different configurations are there now?

4. If you are familiar with the instant insanity cubes, consider the instant insanity problem. The problem is solved when each of the four rows of faces contains exactly one face of each color. If each of the four cubes can be rotated in any direction, how many different possibilities are there from which to choose the solution? If you randomly color each side of each cube, what is the likelihood of finding a solution?

5. In the computer networking problem, it is noted that it may be cheaper to send a message from computer A to computer C by using computer B as an intermediary than to construct a direct communications link from A to C. Give an example of a situation where that would be true. What factors contribute to the "cost" of connecting two computers? Are there costs involved in always routing communications between two computers through a third?

Figure 0.9

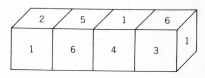

Figure 0.10

CHAPTER 1

FORMAL SYSTEMS

BASIC ASSUMPTIONS

In order to discuss anything about discrete structures, it is essential that we lay down a framework. We will assume only knowledge of high school algebra and geometry, and this chapter will introduce terminology and techniques necessary to explore the remainder of the book. Because one of the goals in learning about discrete structures is to be able to convince someone else of the validity of your arguments and algorithms, we must examine what it means for something to be "true," "correct," "valid", or "right" in a precise way.

SETS AND ELEMENTS

Precision is essential in dealing with discrete structures. Thus, let us be sure that we are all speaking the same scientific or mathematical language. We begin with the blocks with which we will build our discrete structures: **sets.** A set is simply a collection of things called **elements** or **members.** At first, we will consider only those sets that are accompanied by a rule for determination of membership in that set. For any possible element, you must be able to test it with the rule and be able to tell that the element does or does not belong in the set in question. This is not always easy. As an example, the mathematician Fermat claimed that for no positive integers x, y, z, and N (where $N \geq 3$) can we have

$$x^N + y^N = z^N$$

Suppose we want to find the set of positive integers x, y, and z which *do* satisfy this rule. For large values of N, this test can be practically impossible.

How does a rule defining a set look? It can be written out in any way, as long as it clearly defines all the elements of the set. For example, consider the set of all authors of this textbook. If we give that set a name for convenience and call it set Q , then we can write

$$Q = \{ \text{ all authors of this textbook } \}$$

The braces are commonly used to surround the definition of set membership in this fashion. Alternately, we can list all of the elements of the set:

$$Q = \{ \text{ David W. Straight, Shari Lawrence Pfleeger } \}$$

Clearly, this last approach can be useful only when we are dealing with small sets, for it is certainly awkward for very large sets and impossible for infinite sets. (Intuitively, we say that a set is **finite** if there is a positive integer equal to the number of elements in the set. Thus, a set is **infinite** if it has at least as many elements as the set of integers. This means that there is no "stopping point integer" such that the set has only that many elements in it.) Does either of these definitions comply with the restriction placed above, namely, that given any possible element, you can tell whether that element is in the set or out of it? Yes, because we can answer the questions

> Is this candidate element an author of this textbook?

or

> Is this candidate element either David W. Straight or Shari Lawrence Pfleeger?

It is important to note that the order of the elements in the set does not matter. The set

> { Shari Lawrence Pfleeger, David W. Straight }

is the same as the set

> { David W. Straight, Shari Lawrence Pfleeger }

SET NOTATION

To denote membership in a set, we use a symbol as a shorthand. If element a is a member of set Z, we write

$$a \in Z$$

If a is not a member of the set Z, then we write this symbol with a bar through it:

$a \notin Z$

We can use this symbol in defining the set, too. For instance, let I be the set of all integers. Then, we can define the set of all even integers by writing

$$E = \{ \ x \in I \mid x = 2 \times y \text{ for some } y \in I \ \}$$

Here, the symbol $\mid$ stands for "such that." This definition says that we are looking for all integers x such that x is a multiple of 2. Equivalently,

$$E = \{ \ x \in I \mid x \text{ is divisible by 2 with no remainder } \}$$

or

$$E = \{ \ ..., -6, -4, -2, 0, 2, 4, 6, ... \ \}$$

In this last example, we are enumerating all of the even integers. The three dots, known as an ellipsis, indicate that the elements continue in the same fashion. These three dots are commonly used, but it is advisable to use something else when defining infinite sets or even large finite ones, because using an ellipsis leaves room for ambiguity. How can we tell what this set is defining?

$$M = \{ \ 1, 2, ... \ \}$$

Set M could be the set of positive integers, the set of prime numbers, the set of powers of 2, or one of a wide variety of other sets. If the definition is ambiguous, then it cannot serve to define a set.

SET DEFINITION

To emphasize how clear the definition must be, consider the following:

$$G = \{ \ \text{barbers who shave all and only those barbers who don't} \\ \text{shave themselves } \}$$

For this example, let us assume that there are no female barbers and that all men shave. Suppose we are approached by Fred the barber, who wants to know if he is a member of this set G. If Fred is an element of G, then Fred shaves only those barbers who don't shave themselves. Who shaves Fred? If Fred shaves himself, then he is a barber who shaves himself; since he is in G, he thus can't shave himself — a contradiction! On the other hand, if Fred doesn't shave himself, then Fred is a barber who doesn't shave himself; therefore, Fred must shave himself — a contradiction! Using this definition of set membership, then, we find that assuming that he is a member implies that he can't be a member of G. This definition of set G somehow cannot serve as a set definition, because a set definition must allow us to tell whether an element clearly is or is not a member. While at first reading the definition may sound clear, further examination reveals serious problems. Exercise 8 shows how it is possible to have a much tighter, more formally defined set and still have problems.

What about the following definition?

$$L = \{\ x \in I \mid x \text{ is odd and } x \text{ is even }\}$$

Unlike set definition G above, this clearly defines a set. For any integer x, we can test to see whether x is both even and odd. Since no integer can be both even and odd at the same time, then no integer will ever be placed in the set L; that is, the set L is empty. An empty set is perfectly legal in set terminology. In fact, it occurs often enough to warrant its own symbol, ϕ, and its own name. The **empty set** or **null set** is the set that contains no elements. If you think of a set as a box, and the set definition as a rule which determines what goes in the box and what stays out, then the empty set is just an empty box. Consider the set

$$Z = \{\ x \mid x \text{ is a } 208-\text{year}-\text{old barber with one green eye }\}$$

Is the set Z identical to the set L? Even though L is a set of numbers and Z a set of barbers, (we presume) both sets are empty. There is only one empty set, that is, the empty set is unique, so L is the same set as Z.

SUBSETS

Just as boxes can be nested one inside another, so too can sets contain other sets as members. Let us look at some examples.

$$A = \{\ a, b, c, d, e, f\ \}$$
$$B = \{\ A\ \}$$
$$C = \{\ a, c, f\ \}$$
$$D = \{\ a, \{\ c\ \}, \phi\ \}$$
$$E = \{\ \}$$
$$F = \{\ a, a, a\ \}$$
$$G = \{\ a\ \}$$

The set A is a set containing six elements. Remember that the ordering of elements of A doesn't matter. Further, each element is counted only once. So, for example, set F contains only one element, namely a. Such a set is called a **singleton** set because it has but one element. G and F are the same set because the elements of G and the elements of F are identical. Set B is like a box within a box. This set contains as its only element the set A. If you are having trouble picturing this, think of B as a big box. Inside this box is another box, and inside the second box are all of the elements of the set A. Similarly, set E is a big empty box. Set E is the null set; this is another way of writing the null set.

Note that all of the elements of C are also elements of A. We say that one set is a **subset** of another if all of the elements of the first are also elements of the second. So, we say that C is a subset of A. Is A a subset of itself? Yes, because every element of A is an element of A. Indeed, since by default every element of the null set belongs to A, we can say that the null set is a subset of A. Thus, it is clear that every nonempty set has at least two different subsets: itself and the empty set.

If a set K is a subset of a set M, then we also say that M is a **superset** of K. We say that a set K is a **proper subset** of a set M if K is a subset of M and is neither the empty set nor the set M itself. This means that there must be an element x which is in the set M but which is not in K. The notation for K being a subset of M is

$$K \subseteq M$$

If K is a proper subset of M, then we leave out the bar below:

$$K \subset M$$

Let us consider the example illustrated by Figure 1.1. The small circle represents the set K and the larger circle, the set M. The depiction of K's

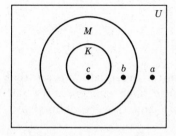

Figure 1.1

circle inside M's circle is a way of showing that K is a subset of M, or, equivalently, that M is a superset of K. The dots represent the elements a, b, and c. The box labeled U represents a larger set containing both K and M. Thus, the element a is an element of U but not of M. Similarly, b belongs to M but not to K. Finally, c belongs to all three of the sets, U, M, and K.

Consider now the set we defined previously as

$$D = \{\ a,\ \{\ c\ \},\ \phi\ \}.$$

This set has three elements, two of which are sets themselves. Thinking again of D as a large box, it contains three items: the lone element a, a small box containing the element c, and an empty box. It is important to see that the second element mentioned, $\{\ c\ \}$, is different from the element c itself. The set $\{\ c\ \}$ is a box containing the element c. So, c is not an element of D, but $\{\ c\ \}$ is an element of D. Likewise, $\{\ a\ \}$ is a subset of D, but $\{\ a\ \}$ is not an element of D.

POWER SETS

The number of subsets of a set is an important concept of discrete structures. It is often the case that we define a set having a certain property, and we then have to enumerate each of the possible subsets of that set. The set of subsets of a set K is called the **power set** of K. If Y is the set $\{\ r,\ s,\ t\ \}$, then the power set of Y has the following elements:

$$\{r\} \qquad \{s\} \qquad \{t\}$$
$$\{r,s,\} \qquad \{r,t,\} \qquad \{s,t\}$$
$$\{r,s,t\} \qquad \phi$$

Exercise 7 will show you that if a set has n elements, then its power set has 2^n elements.

EQUALITY OF SETS

When are two sets identical or equal? We say that two sets are identical when they have exactly the same elements. We saw in the previous examples that when the sets F and G were defined as

$$F = \{\ a,\ a,\ a\ \}$$

and

$$G = \{\ a\ \}$$

we said they were equal because each contained only the element a. Let us examine this more closely. Since G contains the single element a, we see that

$$G \subseteq F$$

On the other hand, even though the element a is shown in the definition of F three times, it is still the only element of F. Thus, we can also say that

$$F \subseteq G.$$

More formally, we say that set A equals set B if and only if the following two conditions hold:

$$A \subseteq B$$

and

$$B \subseteq A$$

This says that every element of the set A is an element of the set B, and also every element of the set B is an element of the set A. It is not always as easy to demonstrate set equality as you may think. For example, suppose A is the set of all standard Pascal programs. Let B be the set of all programs accepted as correct by our Pascal compiler. Are A and B identical? If so, we have established the correctness of our compiler. How can we show set identity here? Can we show that A is a subset of B? That would involve showing that every standard Pascal program is accepted as correct by our Pascal compiler. Then, to show that B is a subset of A, we would have to show that any program accepted as correct by our compiler is a standard Pascal program. What we have accomplished is to describe formally the way to break this problem into two smaller problems in order to solve the larger.

VENN DIAGRAMS

Let us now examine what is outside a set as well as in. If the **universe** is the set of all possible elements of any possible set, then we can consider not only what is *in* a set but also what is *not* in a set. If we have defined a set K,

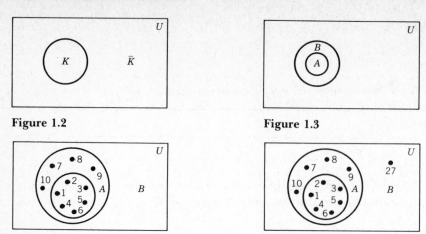

Figure 1.2

Figure 1.3

Figure 1.4

Figure 1.5

then the **complement** of K is the set of all things in the universe that are not in K. We can represent this in a drawing called a **Venn diagram.** In such a drawing, a large rectangle is used to represent the universe, and circles inside the rectangle represent sets. Thus, if the circle in Figure 1.2 represents K, then the rest of the space in the rectangle but outside of the circle represents the complement of K. We denote the complement of K as $\bar{K}$.

Subsets can be represented in a Venn diagram, too. In Figure 1.3, A is a subset of B, so we can place a circle in the universe, label it B, and place another circle inside the first one to represent A. If A is a proper subset of B, then A's circle will be contained in B's circle.

In fact, we can even place the elements of A and B inside their respective circles. Suppose, for example, that A is the set { 1, 2, 3, 4, 5, 6 } and B is { 1, 2, 3, 4, 5, 6, 7, 8, 9, 10 }. We can represent their relationship as a Venn diagram in Figure 1.4. In Figure 1.5, the element 27, which is in the universe but in neither A nor B, can be placed inside the rectangle but outside of either of the two circles. In this manner, we can represent elements that are in the complement of one or both of the sets in question.

FORMAL SYSTEMS

Now that we have sets and elements to work with, we want to examine their properties. How do we start? We have to determine what we know and what we want to know. In computer science, formal systems can be used to do this. They can be useful in establishing program validity and the equivalence of different approaches to the same problem, and in designing and operating the logic components of a computer. To define what is known as an **axiomatic** or **formal system,** we must define each of its five parts:

> primitives
> postulates
> definitions
> rules of derivation
> theorems

Let us look at a formal system with which most of us are familiar to see what its parts are. The legal system in the United States is a formal system. It is generated by terms and concepts whose meanings were agreed upon ahead of time. The U.S. Constitution is the foundation of this legal system. The constitution contains formal, descriptive sentences about concepts that we all understand but never define. For example, the constitution guarantees us certain rights. However, nowhere does it define terms such as "right" or even "freedom." These undefined terms are some of the *primitive terms* in this formal system. The descriptive sentences that are embodied in the constitution involve the primitive terms and their relationships to one another. In our formal system, these are known as *postulates.* The laws of the country, then, consist of *definitions* of terms and conclusions that logically follow from these postulates or from previous conclusions. The logic which the courts and the government use to interpret and generate new laws of the country is the set of *rules of derivation,* that is, the allowable ways in which existing laws and definitions may be extended to new laws. It is important to see that the constitution is a document that can be changed. Although it is the basis for the legal system, the consequences that follow from it are dependent on the assumptions made at the beginning by its writers. If the constitution were to change, so, too, would the conclusions, that is, the laws. Further, a change in the assumptions can have important and often unanticipated changes in the conclusions. Witness the profound effect of the fourteenth amendment to the U.S. Constitution. Originally intended to free the slaves, it has become the basis of a large variety of civil rights decisions that are not related directly to the slavery issue.

In a similar way, to establish a formal system, we must have each of the following:

1. **Primitives:** undefined or intuitive terms.
2. **Postulates** or **axioms:** descriptive sentences about the primitives.
3. **Definitions:** items defined in terms of the primitives and previous definitions.

4. **Rules of derivation:** rules for creating new statements about the system from primitives, definitions, and existing statements about the system
5. **Theorems:** statements logically derived from the postulates and from previous theorems using the rules of derivation

A simple formal system with which you may be familiar is that of plane geometry. The primitives include such things as points, lines and planes. Note that primitives can also include relationships and operations. For example, we say that a point is between two other points, but we never define the relationship of "betweenness." The postulates are assumptions we make or notions we have about the primitives. We say that a straight line can be drawn between two points, or that equal things subtracted from equal things are equals, but we do not prove these statements in any way. We define new things in terms of the primitives: rectangles, equilateral triangles, and the like are defined in terms of points and lines. Finally, using a logical set of rules of derivation, our theorems are derived from the other elements of this formal system. So, when we say that the area of a square is equal to the square of the length of a side, it is because we can derive that result based on the formal system of plane geometry. Again, please note that such a result might be different in a formal system with different axioms and primitives.

TRUTH

If we are to use discrete structures in a meaningful way (that is, to apply what we know is true in order to derive the truth of other things), then it is important that we view discrete structures in the context of a formal system. As we study the different disciplines that comprise discrete structures, we will be introducing definitions of things, concepts, relationships, and operations. We will attempt to use these structures to understand more about what we already know and to expand the horizons of our knowledge. Thus, beyond the formal system in which discrete structures are couched, we must have some idea of what "truth" really is. We need to know what a "valid" argument is and what it means for one result to follow from a set of others. Unless we understand truth and validity, we cannot expand our system of discrete structures with theorems. In other words, we would have definitions and relationships, but we could have no results and could say nothing new about the way the discrete structures interact.

What do we mean when we say that something is true? Consider the following statement:

Chuck has green eyes.

Somehow, we know that this statement is either true or false (i.e., not true). Further, we know that it cannot be both true and false at the same time. However, the phrase,

What time is it?

is neither true nor false. The first sentence is a proposition describing a condition. The second statement is not. In general, when presented with a proposition describing a condition, we can assign it one of two **truth values.** The set of possible truth values in our (conventional) truth system is { *true, false* }. Thus, we say that a statement is true if and only if it is assigned a truth value of *true.* You will see in the exercises that the set of truth values can be different from the conventional set. Indeed, it can have more than two elements. (See Exercise 23.) However, for the purposes of this book, we will consider *true* and *false* the only possibilities. It is important to see that, since *true* and *false* are the only possible truth values, each is the complement of the other. In other words, if a sentence is capable of having a truth value assigned to it and the sentence is not true, then it must be false. Conversely, any sentence which is not false must be true.

PROPOSITIONS

In a formal system that we investigate, we will be considering only **propositions,** that is, statements that can be assigned a truth value. Let us reflect on *when* we assign that truth value. It is important to realize that a statement is usually made in *context.* As long as the context remains the same, the truth value of the statement should remain the same. If the context is unclear, then the truth value of the statement may be unclear as well. For example, the statement

Ben is five years old.

is true or false, depending on which Ben we are talking about and what year it is. Statements such as

It snowed in Chicago on January 16, 1984.

are much easier to assign a truth value, since their truth is independent of their context. However, the statement

NUM < 100

requires a context before any truth value can be assigned. Since time can be part of the context of a proposition, the truth value of a proposition is assigned for the moment, and this value must be reevaluated whenever there is a possibility that the truth value may have changed. Most propositions about everyday life must therefore be reexamined. Statements may even seem to be unchanging but in fact may change their truth value. For example, the statement

Chuck has green eyes.

seems to have a permanent truth value, but it may be the case that Chuck has a glass eye whose color can be changed by inserting a new one at his whim. In the same way, when you are evaluating the truth of a statement in a program, the truth value must be dependent on that particular instant in time. If your program says

```
REPEAT
     NUMBER:= NUMBER + 1
     UNTIL (NUMBER > 5);
WRITELN ('END OF PROGRAM')
```

then the proposition NUMBER > 5 must be evaluated several times for its truth value. Clearly, the second statement changes the value of NUMBER, so the truth value of NUMBER > 5 will change accordingly. On another level, two programs may be trying to access the same data item at the same time. One program wants to change Joe's grade point average at the very same time that a second program wants to read Joe's grade point average. The value of the grade point average may determine which logical sets of operations are performed, and so the truth value involved in a statement such as

DO WHILE (GRADE POINT AVG IS GREATER THAN 2.0)

may be dependent on the time at which the data are accessed. That is why many data base management systems include a way to lock a subset of the data base so that the truth values of your statements don't change as you manipulate and test the data in question.

COMBINING PROPOSITIONS

Let us look at combinations of propositions. Suppose we have two propositions, each of which has a truth value. How can we combine these propositions to make new propositions? If we have

Chuck has green eyes.

and

      Janet has brown hair.

then the following are possible combinations:

      Chuck has green eyes and Janet has brown hair.

and

      Chuck has green eyes or Janet has brown hair.

The *and* and *or* operations act to connect two statements to make new ones. We sometimes call AND a **conjunction** of two statements, and OR is referred to as a **disjunction.** Symbolically, we can let P and Q be two propositions. We want to know the truth values of the resulting statements

    P AND Q

    P OR Q

To do this, we set up what is known as a truth table. A **truth table** is a display of all of the truth values of the simple propositions and the resultant values of the compound proposition. Since P can be either true or false and likewise for Q, the truth table for P AND Q looks like Table 1.1. This is a tabular way of saying that the conjunction P AND Q is true only when both P and Q are true at the same time. Try to verify this with statements about things with which you are familiar.

    Similarly, given the propositions P and Q, we can build a truth table as in Table 1.2 for the disjunction P OR Q. In this case, the disjunction is true whenever at least one of P and Q is true.

    For example, if we know that Chuck has green eyes but Janet does not have brown hair, the statement

      Chuck has green eyes or Janet has brown hair.

P	Q	P AND Q
T	T	T
T	F	F
F	T	F
F	F	F

Table 1.1

P	Q	P OR Q
T	T	T
T	F	T
F	T	T
F	F	F

Table 1.2

is still true. The only time the statement would not be true is when Chuck does not have green eyes and Janet does not have brown hair.

NEGATION

We can also look at a truth table for the proposition NOT P. As you recall, since a proposition can be either true or false, then the proposition NOT P represents the proposition of opposite truth. So, if a proposition P is true, then NOT P is false. Likewise, if NOT P is true, then P is false. We depict that in a truth table as Table 1.3.

Let us pause in our discussion of truth tables to examine what it means to *negate* a proposition. Going back to our previous examples, let P be the proposition

Chuck has green eyes.

We can write the negation of P, namely, NOT P, in several ways:

Chuck does not have green eyes.

or

It is not the case that Chuck has green eyes.

What about the following proposition?

Chuck always eats cookies.

To negate this, we have to write the statement that would make this not true. One choice is

It is not the case that Chuck always eats cookies.

To what simpler statement is this equivalent? You may think that we can write one or some or all of these:

Chuck never eats cookies.

or

Chuck sometimes eats cookies.

or

P	NOT P
T	F
F	T

Table 1.3

Chuck sometimes doesn't eat cookies.

or

Chuck does not eat cookies.

or

Chuck is on a diet.

Clearly, these are not all equivalent. How do we choose the correct one(s)? Let's think about what the words "all" and "always" mean. If, for example, a product claims to be "all natural," what would it take to make that claim false? It would take only a single ingredient's being nonnatural to dispute that claim. Hence, if the proposition "all ingredients in this product are natural" is false, then the statement "one or more ingredients in this product are not natural" or "some ingredients in this product are not natural" must be true. In the same way, the negation of "Chuck always eats cookies" is "Chuck sometimes does not eat cookies."

In a similar fashion, consider the proposition

Emma never rides her bicycle.

The negation of this is

Emma sometimes rides her bicycle.

It is important to see that "all" and "none" are not negations of one another; neither are "always" and "never." The exercises at the end of this chapter will give you a chance to test out your skill at negating propositions.

Let us examine what happens when we use the AND and OR connectors with one proposition, P. That is, we want to form the new propositions P AND P and P OR P. Just as we constructed truth tables for similar combinations of propositions, we can construct truth tables for these: Tables 1.4 and 1.5.

P	P AND P
T	T
F	F

Table 1.4

P	P OR P
T	T
F	F

Table 1.5

P	NOT P	P AND NOT P		P	NOT P	P OR NOT P
T	F	F		T	F	T
F	T	F		F	T	T

Table 1.6   **Table 1.7**

In a similar fashion, we can look at P and its negation in Table 1.6.

Using the OR, we have Table 1.7. Note the last column of this last case. No matter what the value of P, the statement P OR NOT P is always true. This is an example of what is known as a tautology. In general, a statement is a **tautology** if it is true no matter what the truth values of the propositions of which it is composed. In the exercises, we will see other examples of tautologies. Tautologies are important in computing because they can result in code that is always executed, no matter what the circumstances. For instance, suppose your program contains the test

        IF (ALPHA OR NOT (BETA AND NOT BETA))

Because ALPHA OR NOT (BETA AND NOT BETA) is a tautology, the result of this test will always be true, regardless of the values of ALPHA and BETA, so the ensuing statement will be executed.

Before we continue to discuss compound propositions, let us consider what it means to form some of these compounds. When we use an AND to connect two propositions, we want the truth of the resultant statement to be TRUE only when both of the propositions which make it up are true. However, when we have an OR connecting two statements, we can have the resulting compound true if at least one of the propositions which make it up is true. What happens if both propositions are true? Conventionally, since that satisfies the requirement that at least one is true, we say that the composite is true. However, it is often useful to consider a different kind of OR in which the composite is true if and only if exactly one of the propositions which make it up is true. This is known as an **exclusive or,** since it excludes the possibility of having both propositions being true at the same time. We denote an exclusive or by XOR. Thus, P OR Q is true if and only if P is true or Q is true or both P and Q are true. However, P XOR Q is true if and only if exactly one of P and Q is true; that is, one *but not both* of P and Q is true.

Let us consider an everyday example of the difference between OR and XOR. Suppose you are deciding which courses to take for the upcoming year. You say:

I will take Data Structures OR I will take Systems Programming.

This sentence is true if you take Data Structures, Systems Programming, or both. However, the statement

I will take Data Structures XOR I will take Systems Programming.

is the same as saying that you will take either Data Structures or Systems Programming but not both. In other words, the XOR statement is true only if you take exactly one of the two courses mentioned. This distinction could be important if the two courses are taught at conflicting times or if, because of enrollment limitations, a student may not take both courses.

In programming, one binary string is often logically combined with another binary string in order to produce some result. It is especially useful to have an XOR here. For example, if we consider 0 and 1 to have opposite values (so that 0 is FALSE and 1 is TRUE), then we can set

$$P = 01011100$$

and

$$Q = 11001100$$

Then, NOT P would be 10100011 and NOT Q would be 00110011. P OR Q becomes

$$P \text{ OR } Q = 11011100$$

and P XOR Q becomes

$$P \text{ XOR } Q = 10010000$$

You can see that these logical operations come in handy when trying to merge the information contained in different strings of bits.

The XOR connective is also useful in forming the ones complement of a sequence of 0s and 1s. In the **1s complement** of a string, all of the 0s are changed to 1s and all the 1s to 0s. This can be done by using an XOR with a string made up entirely of 1s. For example, if a byte of information looks like

01011001

we can XOR this byte with

    11111111

to get

    10100110

The result (verify it for yourself!) is the original byte with all the bits changed to their "opposites."

NEGATING COMPOUNDS

Having mastered the art of negating simple propositions and of forming compound ones, let us move on to negating compound ones. Suppose we have a proposition which is the conjunction of two simpler propositions:

    I like lasagna and you like pizza.

How is the negation formed? This is a statement of the form P AND Q, where P corresponds to "I like lasagna" and Q to "you like pizza." The negation of the statement should be a statement which is false when P AND Q is true, and true when P AND Q is false. So, we first must examine when P AND Q is true. The truth table for P AND Q illustrated in Table 1.8 shows us that to make the statement P AND Q false, we must negate P, negate Q, or negate both P and Q.

What does this mean in terms of our example? It means that we negate each of the two simpler propositions and then change the conjunction to a disjunction:

    I like lasagna.

becomes

    I do not like lasagna.

P	Q	P AND Q
T	T	T
T	F	F
F	T	F
F	F	F

Table 1.8

and

> You like pizza.

becomes

> You do not like pizza.

Thus, the full negation of the conjunction

> I like lasagna and you like pizza.

becomes

> I do not like lasagna or you do not like pizza.

More frequently, we can express what we have done here as

> NOT (P AND Q) = (NOT P) OR (NOT Q)

Likewise, we negate OR statements in a similar manner. An OR statement such as P OR Q is true if at least one of P and Q is true. Thus, to negate the statement (i.e., to ensure that P OR Q is false), we must negate both P and Q. In other words,

> NOT (P OR Q) = (NOT P) AND (NOT Q)

These rules of negation are known as **De Morgan's Laws.** This sort of expansion can be useful when you are forming or changing conditions in your programs. For example, you may have an IF statement that looks like

> IF (NUMBEROFCHECKS > CHECKLIMIT) AND (LOWBALANCE ≤ LOWLIMIT)
> THEN CHARGE FEE

For various reasons (for example, to structure your program so that the flow is easier to follow, or to accommodate a change in the bank's policy), you may want to negate the test conditions. In this case, we can apply De Morgan's Laws to get

> NOT ((NUMBEROFCHECKS > CHECKLIMIT) AND (LOWBALANCE ≤
>   LOWLIMIT))
> = (NOT (NUMBEROFCHECKS > CHECKLIMIT)) OR (NOT (LOWBALANCE ≤
> LOWLIMIT)) = (NUMBEROFCHECKS NOT > CHECKLIMIT) OR
>   (LOWBALANCE NOT ≤ LOWLIMIT)
> = (NUMBEROFCHECKS ≤ CHECKLIMIT) AND (LOWBALANCE > LOWLIMIT)

Thus, your new code might read

IF (NUMBEROFCHECKS ≤ CHECKLIMIT) AND (LOWBALANCE > LOWLIMIT)
   THEN GIVE BONUS

The same can be done for disjunctions, that is, OR statements. If we are given the propositions

I will go to the market or you will work in the yard.

then the negation becomes

I will not go to the market and you will not work in the yard.

Likewise, in your program, you may have

IF NOT ((GRADES > 85) OR (SATSCORE ≥ 1200))
   THEN NOADMIT

which is the same as

IF ((GRADES ≤ 85) AND (SATSCORE < 1200))
   THEN NOADMIT

How do we convince ourselves that De Morgan's Laws are true? The easiest way is to use truth tables. Since statements with identical truth tables are the same, we can set up all of the possible truth values for P and Q in Table 1.9 and check that both sides of the equality have the same truth values. Note that the column of truth values for NOT P OR NOT Q is exactly the opposite of the values for the column for NOT (P AND Q). Thus, we have compared all of the possible truth values of P and Q and have demonstrated that, in all cases, the first equality holds. Such a technique is useful for any equality which equates combinations of propositions.

P	Q	P AND Q	NOT P	NOT Q	(NOT P) OR (NOT Q)	NOT(P AND Q)
T	T	T	F	F	F	F
T	F	F	F	T	T	T
F	T	F	T	F	T	T
F	F	F	T	T	T	T

Table 1.9

TESTS FOR LOGICAL EQUALITY

Let's look at this possible equality:

NOT (*P* AND (NOT *Q*)) =? *P* OR *Q*

We can easily test this to see if it is indeed an equality by constructing a truth table for each side and ensuring that both sides of the supposed equality have the same truth value for each set of possibilities for *P* and *Q* . If we find even one instance where the truth values of the respective sides of the equality don't match, then this is not an equality at all. Alternatively, we can use De Morgan's Laws to reduce the left—hand side to a simpler form. The left side then becomes

NOT (*P* AND (NOT *Q*)) = (NOT *P*) OR (NOT (NOT *Q*))

Since we have a two—valued logic system, namely, true or false, we can easily see that NOT (NOT *Q*) is the same as *Q* . So, our simplification of the left—hand side becomes

(NOT *P*) OR *Q*

Constructing a truth table for the resultant

(NOT *P*) OR *Q* =? *P* OR *Q*

we have Table 1.10. By looking down the rightmost two columns, we can compare the truth values for each side of the supposed equality. If the columns match, then the two sides always have matching truth values for given truth values of *P* and *Q* , so it is indeed an equality. However, if there is at least one case where the two sides do not match, then this is not an equality at all. In this example, there are mismatches in the second and fourth rows, so we can conclude that this is not an equality, that is, that NOT (*P* AND NOT *Q*) is not the same as the expression *P* OR *Q* .

On the other hand, constructing a truth table for the supposed equality

P AND ((NOT *P*) OR *Q*) =? *P* AND *Q*

we find in Table 1.11 that the columns representing each side do indeed match up:

P	Q	NOT P	NOT P OR Q	P OR Q
T	T	F	T	T
T	F	F	F	T
F	T	T	T	T
F	F	T	T	F

Table 1.10

P	Q	NOT P	(NOT P) OR Q	P AND ((NOT P) OR Q)	P AND Q
T	T	F	T	T	T
T	F	F	F	F	F
F	T	T	T	F	F
F	F	T	T	F	F

Table 1.11

Thus, we have proved, using truth tables, that this is an equality. Note that showing a proposition *correct* by use of truth tables means that we have to check every possible evaluation of the proposition. We cannot simply look at some of the instances (such as when P is true and Q is true, or when P is true and Q is false) to draw a conclusion about the truth of a proposition. This is because when we say that a proposition is *true*, we mean that a proposition is always true, regardless of the truth values of the propositions it incorporates.

IMPLICATION

Very often, in proving that something is true, we want to verify an implication, namely, that *if* one set of propositions is true, *then* a conclusion must be true. For example, we may say

If $n < k$ is true, then $2 n < k$ is true.

Symbolically, we can represent this as IF P THEN Q . Sometimes, this is also written as $P \rightarrow Q$. Such a statement is called a **conditional statement** or an **implication**; P is called the **hypothesis** or **antecedent,** and Q is called the **conclusion** or **consequent.** If we know the truth values of P and Q , what is the truth value of the conditional statement IF P THEN Q? Let us examine an example to see what makes the most sense. We can say

If David is paid on Friday, then he will go to the movies.

Let us consider the cases. If David does get paid on Friday (i.e., if P is true), and if he does go to the movies (i.e., if Q is also true), then the implication is indeed correct. Suppose that David gets paid on Friday (P is true) but does not go to the movies (Q is false). Then the implication is incorrect, since the consequences do not follow from the hypothesis. What if David does not get paid on Friday (P is false)? We cannot say that the implication is false, whether David goes to the movies or not, because we haven't satisfied the conditions of the implication. Thus, if David does not get paid on Friday, the implication is true (not false) independent of the consequent.

P	Q	IF P THEN Q
T	T	T
T	F	F
F	T	T
F	F	T

Table 1.12

We can summarize this in a truth table for an implication, shown as Table 1.12. It is important to recognize here that if the statement IF P THEN Q is true, this tells us nothing about the truth of P. As we saw in the above example and can see in the truth table, there are instances when P is false but the implication is true, and there are instances when P is true and the implication is true.

In mathematics and computer science, you will hear many different ways of expressing an implication. Implications are often used as the rules of derivation in a formal system. For example, we often talk about necessary and sufficient conditions for things to be true. If we say that "P is a sufficient condition for Q," this is just another way of saying IF P THEN Q. Similarly, "Q is a necessary condition for P" is the same as IF P THEN Q. Other equivalent phrases are:

> Given P then Q .
> Whenever P then Q .
> Q whenever P.
> P only if Q .

Associated with the implication are three other implications. If IF P THEN Q is an implication, then

> IF Q THEN P

is the **converse** of the implication,

> IF NOT P THEN NOT Q

is the **inverse** of the implication,

> IF NOT Q THEN NOT P

is the **contrapositive** of the implication.

Often, you will see the phrase *if and only if* or the abbreviation for it, IFF. For instance, we can say that

Procedure P will terminate if and only if the value of x is zero.

This is a statement which really represents two statements:

If procedure P terminates, then the value of x is zero.

and

If the value of x is zero, then procedure P will terminate.

In general, the proposition

P IFF Q

or, equivalently,

P if and only if Q

represents the two propositions

IF P THEN Q

and

IF Q THEN P

The contrapositive of an implication can be very useful in showing that an implication is true. The contrapositive has the same truth values as the implication, and we show this by forming Table 1.13. It is often easier to prove that the contrapositive is true when trying to prove that the implication is true. If we are given the implication

If the turnaround time on the WARTHOG 4400 is greater than 1 hour, then more than three programs must be running concurrently.

P	Q	NOT P	NOT Q	IF P THEN Q	IF NOT Q THEN NOT P
T	T	F	F	T	T
T	F	F	T	F	F
F	T	T	F	T	T
F	F	T	T	T	T

Table 1.13

P	Q	NOT P	(NOT P) OR Q
T	T	F	T
T	F	F	F
F	T	T	T
F	F	T	T

Table 1.14

then the contrapositive is

> If three or fewer programs are running concurrently on the
> WARTHOG 4400, then turnaround time must be less than or equal
> to 1 hour.

You can see that in this case the contrapositive is much more easily proved true (by testing the processor) than the original implication.

We can, for convenience, express implications in terms of AND, NOT, and OR. This allows us to reduce the number of connectives that we need in forming new propositions. Let us examine Table 1.14, the truth table for the expression (NOT P) OR Q . Comparing this with the truth table for IF P THEN Q, it is easy to see that the values are the same, that is, that

IF P THEN Q = (NOT P) OR Q

This fact can come in handy when trying to prove or disprove implications. (See Exercise 14.) For example, suppose we want to show that the statement

> If Congress does not pass the bill, then I will not get my student

> loan.

is true. Let P be the statement "Congress does not pass the bill," and let Q be the statement "I will not get my student loan." Then it may be easier to establish the truth of the statement NOT P OR Q , which is

> Either Congress passes the bill or I will not get my student loan.

ARGUMENTS AND PROOF CONSTRUCTION

Suppose we are asked to demonstrate the truth of a proposition. For instance, we are asked to show that the sum of the integers from 1 through n is $n(n + 1)/2$. How can we convince others that a proposition such as this one is true? One way is to use truth tables, as shown above. It is easy to see how exhausting this would be with all but the simplest arguments. Suppose, for instance, that we have a series of assertions or implications, and a final conclusion. How can we tell if this conclusion follows from the assertions? Let us look at an example.

> *P*: I will finish my project by Christmas.
> IF *P* THEN *Q*: If I finish my project by Christmas, then I can take two weeks' vacation.
> IF *Q* THEN *R*: If I take two weeks' vacation, then I can visit my family in South Carolina.
> THEREFORE *R*: I can visit my family in South Carolina.

This is an example of an argument. An **argument** is a list of statements and a conclusion where it is inferred that the conclusion follows logically from the statements preceding it. The set of statements is known as the set of **premises.** In the example above, we have three statements (or premises) and one conclusion. How can we tell if the conclusion follows from the premises? Let us think about what the argument means in a logical sense. If we are given a general argument, we have a set of statements

$$s_1, s_2, s_3, ..., s_n$$

and a conclusion *C*. Asserting that the conclusion follows from the arguments means that the information in statement s_1 AND the information in statement s_2 AND the information in statement s_3 ... AND the information in statement s_n IMPLY the conclusion *C*. Logically, then, we can write this as

> IF (s_1 AND s_2 AND s_3 AND   ... AND s_n) THEN *C*

If this IF ...THEN statement is a tautology, then it is true, regardless of the truth values of the propositions of which it is composed. For this reason, we say that the argument is **valid**, that is, that the conclusion follows logically from the premises. So, to show that our sample argument about Christmas vacation is valid, we need to set up the argument logically and then test it in a truth table. The argument is easily seen to be of the form

> *P* AND (IF *P* THEN *Q*) AND (IF *Q* THEN *R*) THEREFORE *R*

or, equivalently,

> IF (*P* AND (IF *P* THEN *Q*) AND (IF *Q* THEN *R*)) THEN *R*

The truth table for this is a bit long (see Exercise 17), but it demonstrates that this is indeed a tautology; thus, the argument is valid.

There is often a simpler way to validate arguments without truth tables. We can establish by constructing a truth table that if we know that P implies Q and that Q implies R, we can thus conclude that P implies R. This means that the implication is *transitive*. We can use this property to demonstrate the truth of a propostion which is at the end of a string of implications. In general, any argument of the form

> IF P THEN P_1
> IF P_1 THEN P_2
> IF P_2 THEN P_3
>
>   .
>
>   .
>
>   .
>
> IF P_n THEN Q
> P
> THEREFORE Q

can be shown to be a tautology by generating the entire truth table (which gets very large as n gets large) or by using the transitivity of the implications to reduce the string of n implications to just one. Either method will show that the implication is a valid one. Knowing that this sequence of general implications forms a valid argument, we can look at similar arguments and verify whether they are in this form. If they are indeed of this form, then they must be valid. For example, let us look at the following "proof."

> Suppose T is an equilateral triangle.
> Since T is equilateral, it has equal angles.
> Since T has equal angles, and since every triangle has 180 degrees, then T must have three angles of 60 degrees each.
> Therefore, each angle of T has 60 degrees.

By showing that this argument is of the form P, IF P THEN Q, IF Q THEN R, IF R THEN S, THEREFORE S, we have shown that it is a valid argument. By creating a chain of implications like this, we are creating what is known as a **direct proof.**

Sometimes, it is not easy to chain together implications like this. In such cases, we sometimes resort to using the contrapositive as our implication to be proved. This sort of proof is known as an **indirect proof.** Since we have shown that the implication is equivalent in truth value to the truth of the contrapositive, this is a perfectly legal approach. For example, suppose we have the following assertion to prove:

A: If I is the set of all positive integers, then I is infinite.

This is of the form IF P THEN Q . However, it is much easier to look at the contrapositive of A:

CA: If I is not infinite, then I is not the set of all positive integers.

To prove A indirectly, we assume that the consequent of the original assertion is *not* true, and we use that to prove that the antecedent is not true. Thus, suppose I is not infinite. Then, there must be a largest positive integer in I. Call that largest positive integer M. If M is an integer, then we know that we can add M to itself to get another integer, $2 \times M$, which is indeed larger than M. Thus, we have shown that contradicting the consequent leads to a contradiction of the antecedent. In other words, we have shown that CA is true, so A itself must be true.

Another proof technique is commonly used when the theorem to be proved is a statement that is a general result having many regular but distinct cases. This proof technique is called **proof by induction.** To give you an idea of the kind of problem for which it is best suited, let us look at the following part of a program. Suppose that you have a nested loop of the form

```
FOR FIRSTNUMBER:= 1 TO LIMIT DO
    BEGIN
    FOR SECONDNUMBER:= 1 TO FIRSTNUMBER DO
      BEGIN
      WRITELN (FIRSTNUMBER, SECONDNUMBER)
      END
    END
```

You want to know how many lines of print this section of your program will produce. This involves knowing the sum of the numbers from 1 to LIMIT. To see why, note that for each value of FIRSTNUMBER between 1 and LIMIT, the inner loop prints out k lines, where k is equal to FIRST-NUMBER. Thus, when FIRSTNUMBER is 1, this program segment prints one line; when FIRSTNUMBER is equal to 2, two lines are printed. Therefore, the problem is solved if we can prove the following:

For n a positive integer, the sum of all of the integers from 1 to n is $n(n + 1)/2$.

INDUCTION AND PROOF CONSTRUCTION

How can we state this in a way that is a set of regular cases? We can write the assertion as

IF n is a positive integer, THEN $1 + 2 + 3 + ... + n = n(n + 1)/2$.

Then, for each positive integer n, we have a different case of the same general result:

$$1 + 2 + 3 = 6$$
$$1 + 2 + 3 + 4 = 10$$
$$1 + 2 + 3 + 4 + 5 = 15$$

and so on. We can use this property of the problem to prove the assertion.

The regularity of the problem allows us to address the problem in two stages. First, we establish the fact that the theorem is true when n is equal to 1. Second, we show that, for every possible choice of a positive integer K, whenever the theorem is true for n equal to K, it is also true for n equal to $K + 1$. These two intermediate results can be used to prove that the theorem must be true for any positive integer. Let us see how. Suppose F_n stands for the statement of the theorem for a particular integer n. In our example, we can let F_n stand for the equality

$$1 + 2 + 3 + ... + n = n(n + 1)/2$$

For instance, if n is equal to 5, then F_5 stands for the equality

$$1 + 2 + 3 + 4 + 5 = 5(5 + 1)/2$$

Our two stages of proof will establish that

Stage 1: F_1 is true.

and

Stage 2: If F_K is true, then F_{K+1} is true.

Since we know that F_1 is true, then stage two tells us that F_2 must be true. Knowing F_2 is true and again applying our second stage result, we establish that F_3 is true. In this way, we can build a chain of truths which establish the truth of F_1, F_2, F_3, F_4, ..., F_n for any positive integer n.

Let us use this approach to prove the assertion at hand. First, we prove the case where n is equal to 1. Clearly, $F_1 = 1 = 1(1 + 1)/2$, so we have established the truth of F_1. Next, we want to show that if F_K is true, then $F_K + 1$ is true, where K is a positive integer. If F_K is true, then we know that

$$1 + 2 + 3 + ... + K = K(K + 1)/2$$

How can we use this information to establish the truth of $F_{K + 1}$? First, we observe that

$$1 + 2 + 3 + 4 + ... + K + (K + 1)$$

can be written as

$$(1 + 2 + 3 + 4 + ... + K) + K + 1$$

However, the truth of F_K tells us that we now have $1 + 2 + 3 + 4 + ... + K + (K + 1) = K (K + 1)/2 + (K + 1)$

$$= [(K \times K)/2] + [K /2 + 2 K /2] + 2/2$$
$$= (K + 1) [(K + 1) + 1]/2$$

Thus, $F_{K + 1}$ is true. As we saw above, we can use the two stages of this proof technique to conclude that F_n must be true for any positive integer.

How can we generalize this technique? We have a statement that relates to a regular increment. In our example, our statement was a function of n, and n increased regularly from one integer to the next. The inductive proof has several parts:

1. Define the statement and the regularity of the increment.
2. Prove the truth of the statement for an initial step.
3. Assume the truth of the statement for a particular point.
4. Show that this assumption implies that the next increment is true.

Most statements involving integers or a regular progression through some set are good candidates for this kind of proof.

There are some pitfalls, however. Let us look at an example to see where care must be taken in an inductive proof. Suppose we want to prove the following assertion:

Everyone in this room has the same birthday.

To prove this by induction, we can use the number of people in the room as our way of incrementing through the inductive proof. So, as our first step, we have one person in the room. Then, clearly, everyone in the room has the same birthday. Now, suppose we know that for up to K people, if we have K people in this room, then they must have the same birthday. We want to show that the assertion is true for $K + 1$ people in the room. So, suppose we have $K + 1$ people in the room. We send one person (person A) out of the room. That leaves the room with K people in it, all of whom must therefore have the same birthday. If we bring person A back into the room and send out someone else (person B), then the remaining K people in the room must have the same birthday, including person A. Clearly, from the first step of sending out person A, person A, person B, and all of the others in the room must all have the same birthday.

Is our proof correct? Obviously, common sense tells us that there must be something wrong. What happens when exactly two people are in the room and you try to use the argument above? The "sending a person out of the room" argument doesn't work. (Try it!) It is for this reason that you must be very careful that your inductive argument can be carried from the initial case to all of the general cases.

There are really two ways of setting up an inductive proof. Both ways begin by formulating the statement you want to prove as a statement F_n, where each case is dependent on the regularity of the assertion. In both methods of induction, you then show that F_n is true for an initial value of n. It is in how the proof proceeds from here that the methods differ. In the first method, as above, you then assume that F_n is true for the case where n is K, and you show that this implies the truth of F_n for the case where n is $K + 1$. We can write this more formally as

1. Define F_n as a set of cases which depend on a positive integer n.
2. Show that F_n is true for an initial case $n = 1$.
3. Assume that F_n is true for all cases $n = 1, 2, ..., K$, and show that this implies that F_n must be true for case $n = K + 1$.

However, sometimes a different sort of induction is useful. Instead of using the previous case to imply the next case, we want to use the result for *any* of the previous cases to imply the truth of the next case. Let's see how this works with an example. Suppose we want to prove the following assertion:

Every positive integer is the product of a set of prime numbers.

We first define our statement in terms of cases:

If n is a positive integer, then n can be written as the product of a set of prime numbers.

Proving this statement true is easy for the initial cases. Clearly, 1, 2, and 3, being primes themselves, are obviously the products of primes:

$$1 = 1 \times 1$$
$$2 = 2 \times 1$$
$$3 = 3 \times 1$$

Suppose we assume that every integer less than a given integer K can be represented as the product of primes. We want to use this information to prove that K itself can be written as a product of primes. So, let us examine K. If K is itself prime, then $K = K \times 1$, and we are done. However, if K is not prime, then we know that there are at least two integers, a and b, where a and b are less than K, and $K = a \times b$. Since a and b are each less than K, then our assumption that the assertion is true for all integers less than K allows us to write both a and b as the product of primes:

$$a = p_1 \times p_2 \times p_3 \times ... \times p_r$$
$$b = q_1 \times q_2 \times q_3 \times ... \times q_s$$

However, now we can write K as the product of these primes:

$$K = p_1 \times p_2 \times p_3 \times ... \times p_r \times q_1 \times q_2 \times q_3 \times ... \times q_s$$

and our assertion is proved.

What have we done here? We have defined our initial statement F_n as before. However, now we assume F_n to be true for all values of n less than K, and we show that this means that F_n must be true for $n = K$ as well. How does this differ from the previous method? Not only are we not restricted to the case immediately preceding K, but we are not even restricted to integers! You will see in later chapters that this form of induction is extremely useful, especially when dealing with graphs and lattices.

Let us summarize this second method formally, as we have done with the first method:

1. Define F_n as a set of cases that depend on a number n.
2. Show that F_n is true for an initial value of n.
3. Assume that F_n is true for every n less than K, and show that this implies that F_n must be true for the case $n = K$.

The exercises that follow will give you practice in proving some assertions. In particular, you will have a chance to develop inductive hypotheses, that is, those statements that express the general rule that steps through the possibilities.

The methods introduced in this chapter will be used many times in succeeding chapters. You will find that, by understanding the examples in each chapter and by doing the exercises at the end of each chapter, you will soon master these techniques and will be able to apply them to the problems that confront you in your computing career.

EXERCISES

1. Can an element be at the same time both a member and a subset of a set? Why or why not?

2. Does every set have a proper subset? Which do and which do not?

3. Draw a Venn diagram for the following sets:

$A = \{ *, @, +, \$, \% \}$
$B = \{ *, @ \}$
$C = \{ +, \$, \% \}$
$D = \{ *, @, \P \}$

4. Draw a Venn diagram to represent the following sets:

$Q = \{ 1, 3, 5, 8 \}$
$M = \{ 1, 4, 5 \}$
$P = \{ 1, 2, 3, 4, 5, 6, 7, 8 \}$
$R = \{ 3, 5 \}$

5. Is every set describable? That is, does every set have a set definition that you can write down? Are there sets for which no definition can be written down?

6. PROGRAMMING PROBLEM: Write a program which will do the following:

1. Read in a list of the elements of set A.
2. Read in a list of the elements of set B.
3. Read in an arbitrary element and test to see whether it is a member of A or B or neither or both. Print out the result.

7. Show that if a finite set A has n elements in it, then the power set of A has 2^n elements in it. **Hint**: How many different binary numbers are there with n binary digits? Why is this relevant? Chapter 5 discusses similar problems for some infinite sets.

8. Suppose we define the set S to be the set of all sets X such that X is not an element of X. Is S itself an element of S? (This is known as Russell's Paradox.)

9. Set up a system of axioms and rules of derivation in which the statement X OR NOT X is true (a tautology) but is not a theorem.

10. Negate each of the following expressions:
1. Robert will do the laundry or Dwight will take it to the cleaners.
2. Either Robert will do the laundry or Dwight will take it to the cleaners but not both.
3. Both Nancy and Martha are carpenters.
4. Neither Paul nor Allen goes to school.
5. Diane reads the book but Larry does not.
6. If Jane calls Norman or Betty calls Jim, then David will go to the movies.
7. If Pat goes to the meeting and Dave works late, then neither will be home for dinner.

11. A set of logical connectives is called **functionally complete** if one can express all possible other logical connectives using only those connectives in the set. For example, in this chapter, we showed that IF...THEN can be expressed in terms of the set { OR, AND, NOT }. Show how you can produce a connective, DOT, which is true only if either A is true and B is false or if A is false and B is true. Is the set of connectives { OR, DOT } functionally complete? Why or why not?

12. Let us define the **exclusive or** function as being true only if A is true and B is false or if A is false and B is true. Is this the same as the definition of XOR in the chapter? Is { XOR } functionally complete? Define a NOR connective as being true if and only if neither A nor B is true. Further, define a NAND connective to be true only when A is not true and B is not true. Generate the truth tables for NOR and NAND. Is { NOR, NAND } functionally complete? Is { XOR, NAND } functionally complete? Is { NAND } functionally complete?

13. Write a truth table for (NOT P) OR Q . Compare it with the truth table for IF P THEN Q .

14. We know that the implication IF ...THEN can be written in terms of AND, OR, and NOT as (NOT P) OR Q . What statement using AND, OR, or NOT is equivalent to its contrapositive? Its converse? Its inverse?

15. What are the inverse, converse, and contrapositive for each of the following implications?

1. If it rains today, I will stay home and work.
2. Only if Jack calls will Fred finish the chapter.
3. Whenever I feel afraid, I whistle a happy tune.
4. Given an ample salary, John will work overtime.
5. Congress will pass the bill only if it is approved by the committee or if the president supports it.

16. Derive the truth table for each of the following expressions:

1. NOT $(P$ OR NOT $Q)$
2. P AND (NOT S) AND (NOT T)
3. IF $(P$ OR $Q)$ THEN Q
4. P OR $(P$ AND $Q)$
5. Q XOR $(P$ AND $Q)$

17. Derive the truth table for

IF [P AND (IF P THEN Q) AND (IF Q THEN R)] THEN R

Simplify this expression using only AND, OR, and NOT.

18. Is the expression

P OR $(Q$ OR $R)$

the same as the expression

$(P$ OR $Q)$ OR R?

In other words, does the grouping matter? (Hint: Test to see if the equality of these two expressions is a tautology.) Does the grouping matter in the following expressions?

IF (IF P THEN Q) THEN R
P AND $(Q$ AND $R)$

Is IF P THEN (IF P THEN Q) a tautology? Is IF P THEN (IF Q THEN P) a tautology?

19. Data on students in the computer science program are represented by 32 bits of a word. The rightmost bit is bit 1. Bit 19 indicates that a student has taken the discrete structures course: the bit is 0 if the course hasn't been taken, 1 if it has. Many programs want to isolate the information contained in this bit, but some machines do not have bit—addressability (i.e., they can access one byte or one full word of information but not one bit at a time). Assume that you can fill a byte or a word with any combination of 0s and 1s but that you can test only for a result which is either all 0s or all 1s. Using what you know about logical operations on a string of bits, how can the information be accessed by

1. Shifting bits?
2. Using the AND connective with another bit string?
3. Using the OR connective with another bit string?

How can this nineteenth bit be set to 1 using connectives? How can it be set to 0 using connectives? How can the bit be flipped, that is, changed to the value opposite to its current value?

20. In some machines, the ones complement of a number n is the bit representation of the number $-n$. The ones complement means that whenever there is a 0 in the representation of n, there is a 1 in the representation of $-n$. Similarly, if there is a 1 in the representation of n, there is a 0 in the representation of $-n$. How can the connective XOR be used to get the ones complement of a given number?

21. PROGRAMMING PROBLEM: Suppose the Computer Science department associates with each student an eight−bit sequence that indicates those required computer science courses that the student has taken. Thus, the sequence

$$b_7 \; b_6 \; b_5 \; b_4 \; b_3 \; b_2 \; b_1 \; b_0$$

represents:

b_0: Introduction to PASCAL
b_1: Advanced PASCAL
b_2: Machine Organization
b_3: Discrete Structures
b_4: Numerical Analysis
b_5: Data Structures
b_6: Assembly Language
b_7: Systems Programming

where 1 indicates that a student has taken the course, 0 that the student has not. Write a program which reads in the student name and the associated bit string for up to 30 students. If your programming language cannot work at the bit level, use character strings of 0s and 1s. Allow your program to

1. Read in information on courses dropped and added.
2. List the courses a particular student has had.
3. List the people who have taken a particular course.

P	NOT P
T	F
F	M
M	T

Table 1.15

22. Suppose the set of possible truth values of a system is { TRUE, FALSE, MAYBE }. Suppose further that the truth tables for AND, OR, and NOT look like Tables 1.15, 1.16 and 1.17.
Construct an IF ...THEN truth table consistent with these definitions so that IF P THEN Q is still equivalent to NOT P OR Q .

23. What is the complement of the set of all valid Pascal programs? Is a valid FORTRAN program in this complement?

24. In designing a logic network with various gates, you need an ample supply of each kind of gate: AND, OR, NOT, and XOR. Manufacturing problems may be simplified if fewer different kinds of gates need to be stocked. Since the set { AND, OR, NOT } is functionally complete, you need only AND gates, OR gates, and NOT gates for a network, no matter how complex. Can you get by with just two kinds of gates? Or even one kind of gate?

25. Develop inductive proofs to prove each of the following theorems:
1. THEOREM: The number of ways of arranging n items on a shelf is n ! where $a ! = 1 \times 2 \times 3 \times ... \times a$ is the product of all positive integers less than or equal to a. (Hint: Use induction on n.)
2. THEOREM: The number of ways of arranging n items in a circle is $(n - 1)!$
3. THEOREM: The number of ways of choosing k items from a bag of n distinct items is

P	Q	P AND Q	P	Q	P OR Q
T	T	T	T	T	T
T	F	F	T	F	T
T	M	M	T	M	T
M	T	M	M	T	T
M	F	F	M	F	M
M	M	M	M	M	M
F	T	F	F	T	T
F	F	F	F	M	M
F	M	F	F	F	F

Table 1.16

$$\frac{n\ !}{k!(n\ -\ k)!}$$

4. THEOREM: There is no longest PASCAL program.

26. PROGRAMMING PROBLEM: Write a program which will accept as input a logical expression (such as those in Exercise 16) and which will produce as output the truth table for that logical expression. Use the results of Exercise 16 as test data.

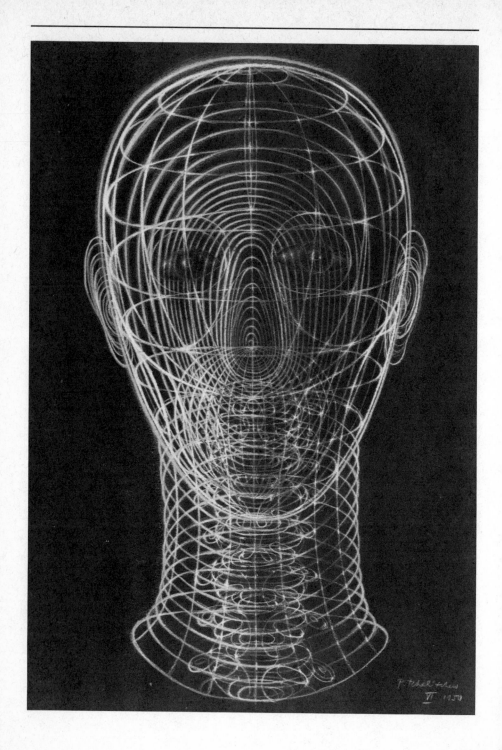

CHAPTER 2

FUNCTIONS AND RELATIONS

In the previous chapter, we spent a great deal of time looking at sets and ways of describing them. In this chapter, we look more closely at collections of sets and the ways in which we can manipulate the elements of sets to form new sets or to give us additional information about the sets. In other words, we will look at relationships among set elements and among sets themselves.

CARTESIAN PRODUCTS

What do we mean by a relationship? All we really mean is that we designate some elements of sets as being related in some fashion; those which are not so designated are considered not to be related. For example, consider the records in the registrar's office at the local university. For each student enrolled, there is a corresponding number of credit hours that the student has completed at the university. The registrar can print out a list of pairs (NAME, CREDITS), where the student's name is the first element of the pair and the number of credits accumulated is the second. This creates a *correspondence* between the set of students and a set of values, and each pair exhibits that correspondence. Note that the order of elements in the pair is important. The first element of the pair must be a member of the set of students enrolled at the university, and the second element is a nonnegative integer. What is the relationship that defines the pair? Simply, it is that NAME has completed CREDITS credit hours at this school. So, for example, if Dennis McCloud has completed 56 credits, then (Dennis McCloud, 56) is a pair defined by the relationship we have described.

How can we describe this formally using set language? If A and B are sets, we define the **Cartesian product of A and B** (denoted $A \times B$) as the set of ordered pairs (a, b) where a is an element of A and b is an element of B. Thus, $A \times B$ contains every possible way we can match up one element of A with one element of B. The order of the elements in the ordered pair is important: (a, b) is an ordered pair in $A \times B$ if a comes from set A and b comes from set B. As you will see in Exercise 1, if A has m elements and B has n elements, then $A \times B$ must have $m \times n$ elements. Let E be the set of students enrolled at the university and N the set of nonnegative integers. Does $E \times N$ represent the relationship we described above? Not quite. $E \times N$ represents all ordered pairs with first element from E and second element from N. Thus, not only does it include (Dennis McCloud, 56) but it includes (Dennis McCloud, n) for every nonnegative integer n. Our relationship is a *subset* of $E \times N$. Let T be the rule that says that NAME has accumulated CREDITS. Then we can define our relationship by saying that we want only those elements of $E \times N$ where the element from E is *related to* the element from N by the rule T. More formally,

$$R = \{ (x, y) \mid x \in E, y \in N, \text{ and } x \text{ is related to } y \text{ by } T \}$$

is the formal relationship we are aiming for.

RELATIONS

In general, then, given sets A and B, a **relation** R is a subset of $A \times B$. Those ordered pairs that are in the subset R are said to be *in* the relation, and we write $(x, y) \in R$ to mean that x is related to y by the relation R. Another way of writing this is $x\ R\ y$. In the example above, for instance, we can write (Dennis McCloud, 56) $\in R$, or (Dennis McCloud) R (56).

Other examples of relations abound. Let us look at the set of programs, and let J be the set of all procedure names. Further, let Q be the set of variable names. An element of the set $J \times Q$ is an ordered pair where the first member of the pair is a procedure name and the second a variable name. We can define a relation R which says that (j, q) is in R if and only if q is a variable required in the procedure call for j. In a similar way, a symbol table is just a relation. If your program has a set M of variable names, and if H is the set of possible addresses of items in your program, then the symbol table is merely that subset of $M \times H$ defined by the following relation: (m, h) is in the symbol table if variable m is located at address h. In other words, m is related to h if and only if variable m is at address h in your program.

In the examples above, we used two different sets in forming the Cartesian product. This does not have to be the case. Look, for instance, at the set Z of integers. We can form the Cartesian product $Z \times Z$ and define relations as subsets of $Z \times Z$. One relation with which you are probably familiar is the one which orders the integers. We say that x is related to y if and only if x is less than or equal to y. Thus, the ordered pairs of the relation look like (x, y) where $x \leq y$. We can write the relation as a subset of $Z \times Z$:

$$R = \{ (x, y) \mid x \text{ is less than or equal to } y \}$$

or we can write out a similar definition

$x \, R \, y$ if and only if x is less than or equal to y

In addition, if a relation is clear from enumerating the pairs in it, we can write out the set of pairs themselves. Since $\{ (x, y) \mid x \leq y \}$ is not a finite set, we will refrain from attempting to write all of the pairs here.

You will see relations described in these several ways as you study different aspects of discrete structures. All of the descriptions are equivalent, and computer scientists use whichever method is the most appropriate to make the material understandable.

We call the relation that we have described above a **binary relation**, because it defines a relation on *pairs* of elements. Actually, there is no need to restrict ourselves to pairs; we can define the same sort of thing on an ordered $n-$tuple, that is, on a set of n ordered elements. For example, suppose n is 3, and define sets as follows:

A is the set of all states in the United States.
B is the set of all senators in the U. S. Senate.
C is the set of all positive integers.

The Cartesian product of these three sets is the set of triples (x, y, z) such that x is an element of A, y is an element of B, and z is an element of C. We can define a relation R as a subset of $A \times B \times C$ in the following way:

$R = \{ (x, y, z) \mid y$ is a senator from state x and received z votes in the last election $\}$

This is called a **ternary relation** because it is a subset of the Cartesian product of three sets. If we have sets A_1, A_2, ..., A_n, we can define the **Cartesian product of** A_1, A_2, ..., A_n as the set of ordered $n-$tuples $(x_1, x_2, ..., x_n)$ where each x_i is an element of set A_i. A **relation** on the sets A_1, A_2, ..., A_n is a subset of the Cartesian product $A_1 \times A_2 \times ... \times A_n$. In fact, we can even look at relations where n is 1. Such a relation is called a **unary relation.** However, unless we state otherwise, we will be dealing only with binary relations, that is, with subsets of Cartesian products of two sets.

Let us examine an important binary relation on sets. Let us say that a set A is related to set B by relation R if A is a proper subset of B. We would write that as $A \, R \, B$ if and only if $A \subseteq B$ but $A \neq B$ and $A \neq \phi$. How do we describe this as a Cartesian product? If C is the collection of all sets, then $C \times C$ is the set of pairs of sets. Our relationship can then be expressed as

$R = \{ (A, B) \mid A$ is a proper subset of $B \}$

PROPERTIES OF BINARY RELATIONS

Relationships can have certain properties. Let us restrict ourselves in this discussion to a binary relation R on a set S. By that, we mean that R is a subset of $S \times S$. We say that a relation R on a set S is **reflexive** if and only if $x \, R \, x$ for every x in the set S. Is this always true for every relation? Clearly, no. The relation "is less than or equal to" on the set of integers is reflexive, but it is easy to see that the relation "is a proper subset of " on the collection of subsets of a set is not reflexive. If $x \, R \, x$ for *no* x in S, we say that the relation is **irreflexive.**

What about reversing the order of the members of the pair? Suppose R is a relation on the set S. We say that R is **symmetric** if (x, y) is in R implies that (y, x) is also in R for every pair (x, y). Is this true for any relations with which you are familiar? Certainly this is not true for "is less than or equal to" (unless x and y are equal). With the "proper subset" relation, if A is a proper subset of B, then the definition of proper subset says that B cannot be a proper subset of A. Hence, $A \, R \, B$ means that B cannot be related to A. Another way of saying this is that (A, B) is in R but (B, A) is not.

For what relation R on a set S can we say that whenever $x \, R \, y$, it is also true that $y \, R \, x$? If S is the set of women and R is the relation "is a sister of," then we have a symmetric relation. Likewise, if S is the set of triangles and R is the relation "is similar to," then the relation is symmetric.

With some relations, it is always true that whenever $x \, R \, y$ and $y \, R \, x$, then $x = y$. For example, if A is a subset of B and B is a subset of A, we can always conclude that A and B are the same set. Such a relation is called **antisymmetric.** A concept similar to antisymmetry is **asymmetry.** Here, if we know that $x \, R \, y$, then we immediately know that y cannot be related to x. Thus, if (x, y) is in R implies that (y, x) is not in R, then R is asymmetric. As an example, think of the relation "is the mother of " on the set of people. If Jane is the mother of Beth, then it can never be the case that Beth is also the mother of Jane.

A third property of relations is called *transitivity.* We say that relation R on a set S is **transitive** if $x \, R \, y$ and $y \, R \, z$ imply that $x \, R \, z$. For the "proper subset" relation, do we have a transitive relation? Yes. If A is a proper subset of B and B is a proper subset of C, then A must, thus, be a proper subset of C. Likewise, the relation "is less than or equal to" is transitive. Is the "is a sister of" relation a transitive relation?

EQUIVALENCE RELATIONS

Consider your home telephone number. In the United States, the number is a ten−digit number, and the digits are grouped in a meaningful way. The first three digits signify the area of the country in which you live, the next three, the area of the city or state in which you live, and the last four identify your particular telephone. We can define a relation as follows:

$x R y$ if and only if x and y have the same area code

It is easy to see that this relation is reflexive, symmetric, and transitive.

A relation that is reflexive, symmetric, and transitive is called an **equivalence relation.** From the other examples described above, it is clear that "has the same number of credit hours as" is an equivalence relation, but "is a proper subset of" is not. You are familiar with many equivalence relations: for example, equality on the set of numbers, or similarity on the set of triangles. Why are equivalence relations special? A relation that is an equivalence relation breaks a set up into equivalence classes. If R is a relation on a set S, then for any element x in S, the **equivalence class** $[x]$ of x is the set of all elements in S that are related to x by R. In set terminology, this is written as

$$[x] = \{ y \in S \mid y R x \}$$

Is every equivalence class nonempty? Yes, because the reflexive property assures us that at least x is in its own equivalence class (with the exception of the relation R which is a subset of $\phi \times \phi$). What do the equivalence classes in the telephone example look like? The relation breaks all of the telephone numbers up into groups, one for each area code. These groups cover the entire set of U.S. telephones, and no two groups overlap.

We can prove two important theorems about equivalence classes:

THEOREM 2.1 ⎯⎯⎯⎯⎯⎯⎯⎯⎯⎯⎯⎯⎯⎯⎯⎯

If R is an equivalence relation on a set S and x and y are elements of S, then

$x R y$ if and only if $[x] = [y]$

To prove this, we must show the implication to be true in both directions. This means that we must show that

1. If $x \, R \, y$, then $[\, x \,] = [\, y \,]$.
2. If $[\, x \,] = [\, y \,]$, then $x \, R \, y$.

First, we assume that x and y are in S and that x is related to y by R. Then, $[\, x \,]$ is the set of all elements in S that are related to x. We want to show set equality, namely, that $[\, x \,] = [\, y \,]$. Recall from Chapter 1 that one way to do this is to show that $[\, x \,]$ is a subset of $[\, y \,]$ and then that $[\, y \,]$ is a subset of $[\, x \,]$. To show that $[\, x \,]$ is a subset of $[\, y \,]$, let a be any element of $[\, x \,]$. Then, by the definition of equivalence class, $a \, R \, x$. We know that $x \, R \, y$, and by transitivity, this means that therefore $a \, R \, y$. Hence, a is in the equivalence class $[\, y \,]$. What have we shown? That every element of $[\, x \,]$ is an element of $[\, y \,]$, so $[\, x \,]$ is a subset of $[\, y \,]$. In a similar fashion (see Exercise 2), it can be shown that $[\, y \,]$ is a subset of $[\, x \,]$, so we have proved that $[\, x \,] = [\, y \,]$. To show the implication in the other direction, now we assume that $[\, x \,] = [\, y \,]$, and we want to show that this means that $x \, R \, y$. If $[\, x \,] = [\, y \,]$, what do we know? We know that x is an element of $[\, x \,]$, but since $[\, x \,]$ and $[\, y \,]$ are the same set, this means that x is an element of $[\, y \,]$. From the definition of $[\, y \,]$, this means that $x \, R \, y$. ■

Another useful theorem tells us that there is no overlap among equivalence classes:

THEOREM 2.2

If $[\, x \,]$ and $[\, y \,]$ are equivalence classes of the relation R on the set S, and if $[\, x \,] \cap [\, y \,] \neq \phi$ (i.e., there is an element s in S which is an element of both $[\, x \,]$ and $[\, y \,]$), then $[\, x \,] = [\, y \,]$.

To prove this theorem, we refer back to Theorem 2.1. If s is an element of $[\, x \,]$, then $s \, R \, x$. Likewise, if s is an element of $[\, y \,]$, then $s \, R \, y$. By symmetry and transitivity, we know that $x \, R \, y$. By Theorem 2.1, this means that $[\, x \,] = [\, y \,]$. ■

These two theorems tell us that equivalence classes break up a set into pieces which do not overlap. Since every element is in some equivalence class, we know that the entire set can be broken up in this manner. This is very useful in many circumstances. For example, suppose we want to try to prove a theorem or examine the consequences of some statement. We may be able to define a relation which is an equivalence relation. By breaking up the problem into nonoverlapping pieces, we can enumerate an exhaustive set of cases to consider. Further, the examination of each case (i.e., of each equivalence class) often is simpler than examining the entire set or problem.

EXAMPLES OF EQUIVALENCE CLASSES

To see how equivalence classes can simplify the manipulation of the elements of a set, let us look at a particular case. Define P to be the set of players in the intramural softball league at the University of Tennessee. We assume that each player is on exactly one team. We can then define a relation R as

$$R = \{ (x, y) \in P \times P \mid x \text{ and } y \text{ are on the same team } \}$$

R is an equivalence relation on P (convince yourself that this is true!), and the equivalence classes determined by R break up P into the set of softball teams. Scheduling the games for the season involves manipulating the equivalence classes rather than dealing with the entire set P itself.

Let us examine in detail another set of equivalence classes. Let Z be the set of integers, and let n be a fixed integer. For any integers x and y, we can define a relation R as follows:

$x R y$ if and only if $(x - y)$ is a multiple of n

To show that this is an equivalence relation, let us examine the three properties required:

1. For x in Z, $x - x$ is $0 = 0 \times n$, so R is reflexive.

2. For x and y in Z, if $x R y$, then $(x - y) = t \times n$ for some t in Z. However, this means that $(y - x) = - t \times n$, so we have $y R x$.

3. For x, y, and z in Z, suppose we know that $x R y$ and $y R z$. Then, $x - y = t \times n$ and $y - z = w \times n$ for some t and w in Z. But now we can rewrite

$$x - z = (x - y) + (y - z) = (t \times n) + (w \times n) = (t + w) \times n$$
which shows that $x R z$.

This relation is known as **congruence modulo** n.

We use congruences every day. For example, suppose we are choosing players for each of four teams. We may have everyone present "count off" from 1 to 4. Then, the "ones" become team 1, the "twos" become team 2, the "threes" become team 3, and the "fours" become team 4. What we have done is assign to each person an integer from 1 to 4 and then established a relation: person A is related to person B if and only if the number assigned to person A is the same as the number assigned to person B. This relation is the equivalence relation congruence modulo 4. The equivalence classes are the four teams. Theorem 2.1 tells us that two people are on the same team only when they have the same number. Theorem 2.2 says that there cannot be a person who is on more than one team; that is, the teams do not overlap. In programming, you will be able to use congruences to consider cases or to assist you in performing tests. For example, suppose you are writing a program which reads in a set of student names and student numbers. The program is to randomly select 10% of the students for an experiment. Such an assignment can be made by adding all the digits of the student number and viewing them modulo 10. What do the equivalence classes look like? Let n be 10. Then, $[1] = \{ x \in Z \mid 1 - x$ is a multiple of 10 $\}$, so

$$[1] = \{ ..., -19, -9, 1, 11, 21, ... \}$$

Likewise,

$$[2] = \{ ..., -18, -8, 2, 12, 22, ... \}$$
$$[3] = \{ ..., -17, -7, 3, 13, 23, ... \}$$
$$[4] = \{ ..., -16, -6, 4, 14, 24, ... \}$$
$$[5] = \{ ..., -15, -5, 5, 15, 25, ... \}$$
$$[6] = \{ ..., -14, -4, 6, 16, 26, ... \}$$
$$[7] = \{ ..., -13, -3, 7, 17, 27, ... \}$$
$$[8] = \{ ..., -12, -2, 8, 18, 28, ... \}$$
$$[9] = \{ ..., -11, -1, 9, 19, 29, ... \}$$
$$[10] = \{ ..., -10, 0, 10, 20, 30, ... \}$$

How do we know that these are the only equivalence classes? Since $11 \in [1]$, we know from Theorem 2.1 that $[11] = [1]$. Likewise, $[12] = [2]$, $[13] = [3]$, and so on.   Examine these classes to convince yourself that every element of Z belongs in exactly one of these ten equivalence classes. In general, for a given n, there are n equivalence classes modulo n. Each class groups together those numbers that have the same remainder when those numbers are divided by n. Exercise 3 will help to convince you that defining the relation in terms of remainders is the same as the congruence definition above.

Although equivalence classes may seem like a strange concept to you, we use them all the time, perhaps unknowingly. For instance, the digits of your Social Security number can be used to define equivalence classes. We can say that two people are related if and only if the first three numbers of their Social Security numbers are the same. The equivalence classes would then break up the population into nonoverlapping groups according to where the Social Security card was issued. In an address book or telephone book, we partition the set of people into classes based on the first letter of their last name. In a dictionary, we partition the set of words into classes based on the first letter of the word. What are the relations that generate these equivalence classes?

REFLEXIVE AND SYMMETRIC CLOSURES

We have seen examples of relations which are not equivalence relations; that is, they are not reflexive, symmetric, and transitive. Sometimes it is useful to examine a given relation and add to it those pairs which would assure that a certain property is satisfied. For example, let the set S be defined as

$$S = \{ 1, 2, 3, 4, 5 \}$$

We define the relation R as

$$R = \{ (1, 2), (3, 4), (1, 5) \}$$

It is clear that R is not reflexive. For instance, if R were reflexive, then $(1, 1)$ would be in R. What pairs must we add to R to make it reflexive? The definition of a reflexive relation tells us that we need

    $(1, 1)$
    $(2, 2)$
    $(3, 3)$
    $(4, 4)$
    $(5, 5)$

We can add these pairs to R to form a new relation R^r which is a reflexive relation containing R. Thus, R^r is the set

$$\{ (1, 1), (1, 2), (2, 2), (1, 5), (3, 3), (3,4), (4, 4), (5, 5) \}$$

We can picture this as Figure 2.1. Let us generalize this process to form a reflexive relation from *any* given relation.

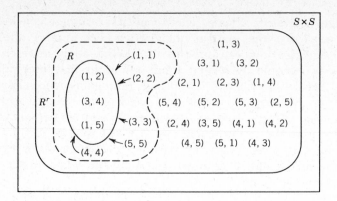

Figure 2.1

If R is a binary relation on a set S, we can form the **closure of R in** S by generating every pair (x, y) such that $x\,R\,y$, with x and y in S. Let us see how this works. Suppose we have a relation R that is not reflexive. If R is a relation on the set S, then R is a subset of $S \times S$. Clearly, since the set $S \times S$ contains the pairs (x, x) for all x in S, the entire set $S \times S$ can be considered a reflexive relation. We could add all of the remaining pairs in $S \times S$ to guarantee a reflexive relation, but we want to add only those that are absolutely necessary to turn R into a reflexive relation. So, somehow, we must be able to add elements to R to form a reflexive relation which contains all of the pairs that are in R. We build such a relation in the following way. We call R^r the **reflexive closure of** R if R^r is a binary relation on S such that R is a subset of R^r and the following is true:

If $a\,R\,b$, then $a\,(R^r)\,a$ and $b\,(R^r)\,b$.

We can easily see that R^r is a reflexive relation. In fact, if R had been reflexive in the first place, then R^r and R would be identical sets.

We can generate a similar new relation from relations that are not symmetric. Suppose R is a relation that is not symmetric. The relation R^r is the **symmetric closure of** R if R^s is a binary relation on S such that R is a subset of R^s and the following is true:

If $a\,R\,b$, then $b\,(R^s)\,a$.

Again, if R is symmetric, then R and R^s are identical. Otherwise, R^s has more elements than R, and the symmetric closure R^s of a relation R is a symmetric relation.

Using the relation we defined on { 1, 2, 3, 4, 5 } above, we see that we would have to add

$$(2, 1), (4, 3), (5, 1)$$

to the original relation to make it symmetric. The symmetric closure of this relation can be pictured as in Figure 2.2.

In building the closure (either symmetric or reflexive) of a relation R, we are adding on whatever is necessary to make the old relation into a new relation with the desired property. We add on exactly what we need in order to have our desired property; in this sense, the closure is the smallest relation having the desired property which contains the original relation. How can we say this more mathematically?

THEOREM 2.3 _____

The reflexive closure of a relation R is the smallest reflexive relation containing R. This means that if T is a reflexive relation containing R, then

$$R \subseteq R^r \subseteq T$$

where R^r is the reflexive closure of R.

_____

The proof of this follows from the definition of R^r and is left as an exercise. ■

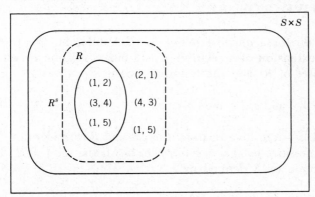

Figure 2.2

Similarly, we have the following:

THEOREM 2.4 _____

> The symmetric closure of a relation R is the smallest symmetric relation containing R. This means that if Q is a symmetric relation containing R, then
>
> $$R \subseteq R^s \subseteq Q$$
>
> where R^s is the symmetric closure of R.

_____

A SPECIAL REFLEXIVE RELATION

There is a reflexive relation which appears often enough to warrant its own name. It is the relation that consists only of elements which are related to themselves. In other words, the **identity relation** I_S on the set S is the subset of $S \times S$ defined by

$$I_S = \{ (x, x) \text{ for } x \in S \}$$

How does this relate to reflexive relations? We know that if the element x of the set S appears in any pair of the reflexive relation R, then the pair (x, x) must also appear in R. Thus, let A be the set of elements of S which appear in any pair of R. Then R is reflexive if I_A is a subset of R. Can we say the reverse, namely, that if I_A is a subset of R, then R is reflexive?

TRANSITIVE CLOSURE

The situation for a *transitive closure* is a bit different. We can define the **transitive extension of a relation** R as a binary relation R^1 on S such that R is a subset of R^1 and the following is true:

> if $a R b$ and $b R c$, then $a R^1 c$.

Again, if R is a transitive relation, then R and R^1 are identical. Let us look at what a transitive extension looks like. If S is the set $\{ 1, 2, 3, 4 \}$, let R be the relation on S defined by

$$\{ (1, 2), (2, 3), (3, 2), (3, 4) \}$$

What does the transitive extension R^1 of R add to this set? Since 1 R 2 and 2 R 3, we must add 1 R^1 3. Likewise, 2 R 3 and 3 R 4 mean that 2 R^1 4. We know that 2 R^1 2 because 2 R 3 and 3 R 2, and we know that 3 R^1 3 because 3 R 2 and 2 R 3. Examine the eight other possible candidates for inclusion in R^1 to convince yourself that we can add no other pairs to R^1 Thus, R^1 is the set

$$\{ (1, 2), (2, 3), (3, 2), (3, 4), (1, 3), (2, 4), (2, 2), (3, 3) \}$$

Is the set R^1 a transitive relation? No. We have (1, 2) and (2, 4) in our new transitive extension relation, but we do not have (1, 4). What we must do is extend R^1 to *its* transitive extension. Define R^2 to be the transitive extension of R^1 and let us see what pairs belong in this new relation. It is a bit tedious, but we examine all pairs of elements of R^1 of the form (x, y) and (y, z) to see whether they generate (x, z) in the transitive extension. We have

$$(1, 2) \text{ and } (2, 3) \rightarrow (1, 3)$$
$$(1, 2) \text{ and } (2, 4) \rightarrow (1, 4) \text{ *}$$
$$(1, 3) \text{ and } (3, 2) \rightarrow (1, 2)$$
$$(1, 3) \text{ and } (3, 4) \rightarrow (1, 4) \text{ *}$$

$$(2, 3) \text{ and } (3, 2) \rightarrow (2, 2)$$
$$(2, 3) \text{ and } (3, 3) \rightarrow (2, 3)$$
$$(2, 3) \text{ and } (3, 4) \rightarrow (2, 4)$$
$$(3, 2) \text{ and } (2, 3) \rightarrow (3, 3)$$
$$(3, 2) \text{ and } (2, 4) \rightarrow (3, 4)$$
$$(3, 2) \text{ and } (2, 2) \rightarrow (3, 2)$$
$$(3, 3) \text{ and } (3, 2) \rightarrow (3, 2)$$
$$(3, 3) \text{ and } (3, 3) \rightarrow (3, 3)$$
$$(3, 3) \text{ and } (3, 4) \rightarrow (3, 4)$$

The comparisons marked with an asterisk indicate those which generated elements not already in R^1 and which must therefore be added to the new transitive extension. Therefore, we have R^2 defined as

$$\{ (1, 2), (2, 3), (3, 2), (3, 4), (1, 3), (2, 4), (2, 2), (3, 3), (1, 4) \}$$

Can we form the transitive extension of this new set?, To form R^3 we would make the same comparisons as above, plus any dealing with (1, 4). Since no pair in R^2 begins with 4 or ends with 1, we have no additional comparisons and thus no additional pairs to add. In other words, if R^3 is the transitive extension of R^2, then $R^2 = R^3$. What about the transitivity of R^2? Examining all of the pairs, you can verify that R^2 is a transitive relation which contains R.

This technique can be generalized in the following way. Let R^{i+1} be the transitive extension of the relation R^i. We define the **transitive closure** of a relation R to be

$$\{ (x, y) \mid (x, y) \in R \text{ or } (x, y) \in R^1 \text{ or } (x, y) \in R^2 \text{ or...or } (x, y) \in R^i \text{ or } (x, y) \in R^{i+1} \text{ or... } \}$$

Thus, we keep extending our original relation until we arrive at a transitive one. How do we know that this process will stop? Since $S \times S$ contains *all* pairs possible and is clearly transitive, we know that we must stop when we reach $S \times S$. How do we know that if we stop short of $S \times S$ that the result will be transitive? The next theorem tells us.

THEOREM 2.5 _____

The transitive closure of a relation R is always a transitive relation.

_____

To prove this, let R^t be the transitive closure of a relation R. If R is transitive, we are done. Let x, y, and z be elements of the set S on which R is a relation, where we have

$$x \, R^t \, y \text{ and } y \, R^t \, z$$

We want to show that $x \, R^t \, z$. Since $x \, R^t \, y$, this means that $x \, R^k \, y$ for some k. Similarly, $y \, R^j \, z$ for some j. If k is greater than or equal to j, then $R^j \subseteq R^k$, so we have $x \, R^{k+1} \, z$. Otherwise, $R^k \subseteq R^j$, so $x \, R^{j+1} \, z$. In either case, since R^t is the union of all of the R^i, we have $x \, R^t \, z$.
■

_____

As with the reflexive and symmetric closures, we can express the transitive closure in terms of its being the smallest transitive relation containing R.

THEOREM 2.6 _____

The transitive closure of a relation R is the smallest transitive relation containing R.

_____

Let R be a subset of $S \times S$, and let R^t be the transitive closure of R. Let Q be a transitive relation containing R. We want to show that $R \subseteq R^t \subseteq Q$. Since R is contained in R^1, R^1 is contained in R^2, and so on, we know that R must be contained in all of the R^i, and thus is in R^t. It remains for us to show that R^t is a subset of Q. Remember that a relation is just a subset of $S \times S$, so we want to show that if (x, y) is an element of R^t, then (x, y) is also an element of Q. If (x, y) is in R^t, then $x R^t y$. This means that $x R^i y$ for some i, so there is a z in S such that $x R^{i-1} z$ and $z R^{i-1} y$. However, Q is a transitive relation containing R, so $x Q^{i-1} z$ and $z Q^{i-1} y$. Because Q is transitive, it follows that $x Q y$. In other words, (x, y) is in Q, and we are done. ∎

EXAMPLES OF TRANSITIVE CLOSURE

To examine another example of transitive closure, let N be the set of nonnegative integers. Define the relation R on $N \times N$ by

$$R = \{ (n, m) \in N \times N \mid m = n + 1 \}$$

What is the transitive closure? Applying the definition, we see that

$$R^t = \{ (n, m) \in N \times N \mid m > n \}$$

How can we use the transitive closure in the real world? If S is the set of all cities in the world, let R be the relation that says that city a is related to city b by R if there exists a direct flight linking city a with city b. The transitive extension R^1 of R is thus a description of how traffic can travel from one city to another on a direct flight or with one stop at an intermediate city. The transitive extension R^2 of R^1 describes the ways in which travel can be made traveling through at most two intermediate cities. Finally, the transitive closure of R shows how air travel can be routed either directly or through any number of intermediate cities.

In computer science, the transitive closure can represent routings in a similar manner. Given a network of computers, message routing can be described by saying that computer A is related to computer B if a message can be sent directly from A to B. The transitive extension of this relation describes the set of computers where messages can be sent between computers using at most one intermediate computer to pass the message. Finally, the transitive closure represents the set of all computers where messages can be sent using any number of intermediate computers; this would represent the entire computer network.

OPERATIONS

Let us turn from relations to a similar concept, that of an *operation*. With relations, we begin with a set S. If we examine two elements of S, we can say either that they are related or they are not. An operation on a set S goes a little farther. Given one or more elements from S, an **operation** produces another element as its result. If an operation requires only one element of S, it is called a **unary** operation. If it requires two elements of S, it is a **binary** operation. If it requires three elements of S, then it is a **ternary** operation. In general, an operation requiring n elements of S is an **n−ary** operation.

Negation is an example of a unary operation that we use all of the time. Given a number x in the set Z, the "−" operation produces the result $-x$ in Z. If we have a set of truth values { TRUE, FALSE }, then the logical negation operator NOT changes TRUE to FALSE and FALSE to TRUE. More often, we deal with binary operations. The arithmetic operations you learned are good examples. The operation "+," for instance, takes two numbers, x and y, in Z and assigns to them a third number, $x + y$. Likewise, "−" as the binary operation of subtraction assigns to x and y the third number $x - y$.

It is important to note the difference between the two operations, negation and subtraction. The former is a unary operation, whereas the latter is a binary one.

The set elements on which the operation operates are known as **operands.** Thus, a unary operation requires only one operand, whereas a binary operation requires two. We can think of an operation, then, as operating on a set of operands to produce a result. It may be useful to think of operations as corresponding to functions or procedures. An n−ary operation is akin to a procedure with n input parameters.

OPERATIONS ON SETS

Let us look at some operations with which you may not be as familiar. These are operations on collections of sets, rather than on collections of numbers. Can you think of a unary operation on a set? Chapter 1 introduced the concept of the complement of a set. "Complement" is a unary operation. It operates on a given set to produce a new set which contains all of the elements not found in the original set. Recall that the Venn diagram for the complement of a set K is as in Figure 2.3, where the shaded section is the complement of K and the complement is denoted by $\bar{K}$.

We can define several binary operations on sets. If A and B are two sets, then we define the **union** of A and B to be the set containing all of those elements that belong to A or B or both. We write the union of A and B as $A \cup B$. More formally, then, we define

$$A \cup B = \{ \ x \mid x \in A \text{ or } x \in B \ \}$$

Using the Venn diagram in Figure 2.4, we can represent the union of A and B with the shaded part. For example, if A is the set of all women with red hair and B is the set of all women with green eyes, then $A \cup B$ is the set of all women with red hair or green eyes (or both). Similarly, if A is the set of PASCAL programs containing at least one CASE statement and B is the set of PASCAL programs containing at least one nested procedure, then $A \cup B$ is the set of PASCAL programs containing a CASE statement, a nested procedure, or both. Exercise 14 will give you a chance to practice forming the union of sets.

Another operation is the intersection of two sets. This operation finds the place where two sets overlap. If A and B are two sets, then the **intersection** of A and B is the set containing all of those elements which belong to A and to B. We write the intersection of A and B as $A \cap B$. The formal definition is

$$A \cap B = \{ \ x \mid x \in A \text{ and } x \in B \ \}$$

The Venn diagram for the intersection is represented by the shaded part where the circles overlap in Figure 2.5.

Again, if A is the set of all women with red hair and B is the set of all

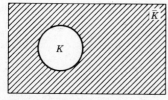

Figure 2.3

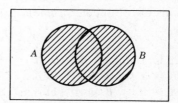

Figure 2.4

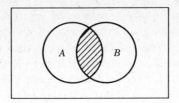

Figure 2.5

women with green eyes, then $A \cap B$ is the set of all women with red hair and green eyes. Alternately, if P is the set of all programs recognized as syntactically correct by the VAX PL/I compiler, and Q is the set of all programs recognized as correct by the IBM 4341 PL/I compiler, then $P \cap Q$ is the set of all programs recognized as correct by both compilers.

Let us examine an example. If P is the set $\{\ 1, 2, 3, 4, 5, 6, 7, 8\ \}$ and Q is the set $\{\ 7, 8, 9, 10\ \}$, then the set $P \cup Q$ is the set of all elements in P or in Q. Thus,

$$P \cup Q = \{\ 1, 2, 3, 4, 5, 6, 7, 8, 9, 10\ \}$$

The intersection of P and Q is the set of elements which are in both sets, so

$$P \cap Q = \{\ 7, 8\ \}$$

What about $R = \{\ 3, 4\ \}$ and its union and intersection with P? Since R is a subset of P, it is easy to see that $R \cup P = P$. Similarly, $R \cap P = R$.

THE POWER SET

Recall that in Chapter 1 we introduced the notion of the power set of a set. If S is a set, then the power set of S, denoted P(S), is the set of all subsets of S. If we consider the operations union, intersection and complement, we see that they operate on elements of the power set of S to produce other elements of the power set of S. What other properties of these operations can we observe? Here is a list of properties that you can verify from the definitions of complement, union and intersection: If A and B are subsets of S, then

1. A is a subset of $A \cup B$.
2. B is a subset of $A \cup B$.
3. $A \cup B = B \cup A$. (This is called *commutativity*.)
4. $A \cup A = A$.

5. If A is a subset of C and B is a subset of C, then $A \cup B$ is a subset of C.
6. $A \cap B$ is a subset of A.
7. $A \cap B$ is a subset of B.
8. $A \cap B = B \cap A$.
9. $A \cap A = A$.
10. If C is a subset of A and C is a subset of B, then C is a subset of $A \cup B$.
11. ϕ is a subset of $A \cup B$ and of $A \cap B$.
12. $A \cup \phi = A$.
13. $A \cap \phi = \phi$.

COMBINING OPERATIONS

With integers, we have operations like addition and subtraction that can be used together. Can we do the same thing with intersection, union, and complement? In other words, what happens when we try to evaluate expressions like

$$A \cup (B \cap C)$$

and

$$A \cap (B \cup C)$$

Let us look at the Venn diagram in Figure 2.6 to see what we are considering. Here we have $B \cap C$. If we now unite A with $B \cap C$, we have the diagram in Figure 2.7. Is this the same as $A \cup B$ intersected with $A \cup C$? Let us look at the Venn diagram for this in Figure 2.8.

It seems to be true. We can prove its truth using the definition of union and of intersection. As with previous examples, we want to prove set

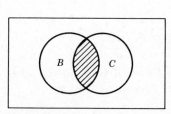

Figure 2.6

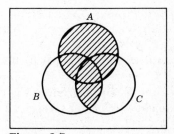

Figure 2.7

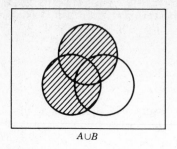 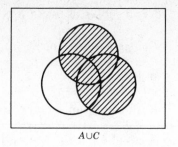

$A \cup B$ $A \cup C$

Figure 2.8

equality, so we show set inclusion in both directions. Thus, first we want to show that the set $A \cup (B \cap C)$ is a subset of $(A \cup B) \cap (A \cup C)$. Suppose x is any element of $A \cup (B \cap C)$. By the definition of union, this means that either x belongs to A or x belongs to $B \cap C$ (or both). If x belongs to A, then by definition of union, it must be in $A \cup B$ and also in $A \cup C$. Hence, by the definition of intersection, x must be in the intersection of $A \cup B$ and $A \cup C$. If x does not belong to A, then it must belong to $B \cap C$. In this case, belonging to $B \cap C$ means that it belongs to B and to C and thus to $A \cup B$ and $A \cup C$. Thus, in either case, x must belong to the set $(A \cup B) \cap (A \cup C)$. The element x was a representative element for *all* elements of the set $A \cup (B \cap C)$, so we have shown that

$$A \cup (B \cap C) \subseteq (A \cup B) \cap (A \cup C)$$

To show set inclusion in the other direction, suppose y is any element of the set $(A \cup B) \cap (A \cup C)$. Then, by definition of intersection, y must belong to both $(A \cup B)$ and to $(A \cup C)$. Since y belongs to $A \cup B$, it must belong to set A, set B, or to both. If y belongs to A, then by definition of union, y must belong to $A \cup (B \cap C)$, and we are done; $(A \cup B) \cap (A \cup C) \subseteq A \cup (B \cap C)$. So, suppose y does not belong to A. Because we know that y belongs to $A \cup B$, this means that y must belong to B. Likewise, y belongs to $A \cup C$, so it must be an element of C. Hence, y belongs to both B and C, so it is a member of $B \cap C$. Finally, this makes y a member of $A \cup (B \cap C)$ and this completes the proof that $(A \cup B) \cap (A \cup C) \subseteq A \cup (B \cap C)$.

Does the same sort of thing work reversing the roles of union and intersection, that is, is it true that

$$A \cap (B \cup C) =? (A \cap B) \cup (A \cap C)?$$

Examine the Venn diagram in Figure 2.9 to see if you can tell whether this is an equality or not. Exercise 15 will help you to decide.

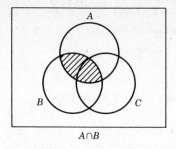 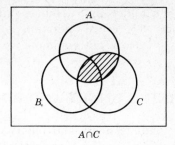

$A \cap B$ $A \cap C$

Figure 2.9

ASSOCIATIVITY

In arithmetic, when we use the same operation on several numbers in succession, we can sometimes perform the operation in different orders and get the same result. For instance, we can add $5 + (9 + 4)$ or $(5 + 9) + 4$, but we get the same answer of 18 each time. We do this by associating the 9 and 4 first and then adding to the 5, or by associating 5 and 9 first and then adding the 4. Since we get the same answer, we say that the operation of addition on the set of numbers is associative. In general, if S is a set, ∇ is a binary operation on that set, and a, b, and c are elements of S, then ∇ is **associative** if

$$a \nabla (b \nabla c) = (a \nabla b) \nabla c$$

for every a, b, and c in S. Is the operation of union on sets an associative operation? Is the intersection an associative operation on sets? Yes, and you can verify this in Exercise 16. Not every operation on sets is associative, however. For example, subtraction on the set of numbers is not associative, since $(3 - 5) - 4$ is not equal to $3 - (5 - 4)$.

COMMUTATIVITY

What about the order in which we perform unions and intersections? Is $A \cup B$ the same as $B \cup A$? In verifying the list of properties following from the definitions of union and intersection, you saw that this was true. The same is true for intersection, namely, that $A \cap B$ is the same as $B \cap A$. In general, if S is a set and ∇ an operation on that set, then ∇ is **commutative** if

$$a \nabla b = b \nabla a$$

for every a and b in S. Hence, intersection and union are commutative operations. Not every operation is commutative. Division on the set of numbers is clearly not commutative.

COMBINING UNARY AND BINARY OPERATIONS

Can we combine unary and binary operations? Just as we can negate the sum of a pair of numbers, so, too, can we take the complement of the union or intersection of sets. In fact, it follows from De Morgan's Laws (see Chapter 1) that the following is true:

THEOREM 2.7 _____

$\overline{(A \cup B)} = \overline{A} \cap \overline{B}$ and $\overline{(A \cap B)} = \overline{A} \cup \overline{B}$.

_____

We will prove the first one and leave the other as an exercise. Suppose x is an element of the complement of $A \cup B$. Then, x belongs neither to A nor to B. Since x cannot belong to A, it must be in the complement of A. Likewise, since x cannot belong to B, it must be in the complement of B. Thus, x must be in $\overline{A} \cap \overline{B}$, and we have shown that $\overline{(A \cup B)}$ is a subset of $\overline{A} \cap \overline{B}$. To show set inclusion in the other direction (and therefore equality), we suppose that y is an element of $\overline{A} \cap \overline{B}$. Then, y must be an element both of $\overline{A}$ and of $\overline{B}$. By definition, then, y is not in A and y is not in B. From De Morgan's Laws, this means that it is not the case that y is in A or B. However, this says that y cannot be in $A \cup B$, and we are done.

■

PARTITIONS AND COVERINGS

The notions of union and intersection are relatively simple ones, but they pervade the study of discrete structures. Many techniques of demonstration and proof employ both these notions and that of an equivalence relation. Let us generalize these to describe a new way of looking at a set. Because both union and intersection are associative, we can consider the union or intersection of large numbers of sets without regard to the way that they are grouped together. For example, we can write

$$\bigcup_{i=1}^{n} A_i = A_1 \cup A_2 \cup ... \cup A_n$$

to represent the union of n sets. Likewise, we write

$$\bigcap_{i=1}^{n} A_i = A_1 \cap A_2 \cap ... \cap A_n$$

to represent the intersection of n sets. What does it mean for an element x to be in $\cup A_i$? Simply that x is in A_i for some i. If y is an element of $\cap A_i$, then y is in A_i for every i. If S is a set where we can write S as the union of a collection of subsets, then we say that the collection **covers** S. In other words, if $\cup A_i = S$, then $\{ A_1, A_2, ..., A_n \}$ is a **covering** of S. Suppose that none of these sets A_i overlaps; that is, there is no element of S which belongs to more than one A_i. We can describe that by saying that $A_i \cap A_j = \phi$ for every pair i and j. The sets A_1, A_2, ..., A_n are said to be **mutually disjoint.** Given a collection of nonempty sets A_1, A_2, ..., A_n which are mutually disjoint and cover a set S, this collection is called a **partition** of S. The sets A_i are called the **blocks** of the partition. If there are n blocks in the partition, the partition is sometimes called an $n-$**partition** of S. As with equivalence classes, partitions are a useful tool, not only in handling discrete structures but also in investigating almost any aspect of computer science. You can partition a problem into distinct cases and handle each individually. Further, if you are counting something, partitions make it easy to count subsets and then sum all the blocks of the partition to find the total you are seeking.

FUNCTIONS

Let us go back to the concept of relation that we described earlier. There, we had two sets and a rule for associating a member of the first with a member of the second. Suppose we picture a "relation machine": a large black box with an "in" tray and an "out" tray. You place an element of the first set in the "in" tray, press a button, and the element disappears. Soon, an element from the second set appears in the "out" tray. You don't really know how the black box works, but you do know that the second element is related to the first by some sort of relational rule. We know this because repeated tests with the same inputs show that the outputs are consistent, not randomized. The output is somehow dependent on the input. Whenever you give it a particular x in a set S, it returns to you the same y in Y, no matter how many times you submit x to the machine. Thus, these ordered pairs produced by the black box form a relation, but this is a very special kind of relation. For each element of S we submit to the machine, we get a *unique* y as a response.

This kind of relation is called a function. In other words, a **function** *f* from the set S to the set Y is a relation which associates to an element of S a unique element of Y. In relational notation, we can write this definition of a function as follows. Let S be a set on which the relation R is defined. If, for every *a*, *b*, and *c* in S, we have

$$(a, b) \text{ in } R \text{ and } (a, c) \text{ in } R \text{ imply that } b = c$$

then R is a function. The subset of S on which *f* is defined is called the **domain** of *f*; the subset of Y to which the elements of the domain are associated by *f* is called the **range** of *f*. Sometimes, we say that *f* **maps** its domain to its range, and we call *f* a **mapping.**

We can denote *f* in many ways. We can write $f:S \rightarrow Y$ to indicate that *f* takes an element in S and assigns it to a unique element in Y. Alternately, we can write $f(A) = B$, where A is the domain of *f* and B is the range. For each element *x* in the domain of *f*, we write $f(x)$ to denote the element in Y to which *f* maps *x*. We call $f(x)$ the **image of** *x* **under** *f* or the **value of** *f* **at** *x*.

Let us examine some functions. If P is the set of programs recognized by the PASCAL compiler on your WARTHOG 7780 computer, and if I is the set of positive integers, then we can define a function *f* which maps a program to the number of entries in its variable list. Clearly, for each program, there is exactly one number of variables, so *f* is indeed a function. Alternately, let H be the set of states in the United States and let I be the set of positive integers. We can define a function *g* which maps a state to its population. For example, g(Tennessee) = 2,000,000 and g(Pennsylvania) = 8,000,000. We can represent the function *g* as a diagram in Figure 2.10.

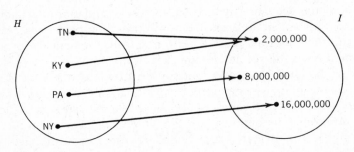

Figure 2.10

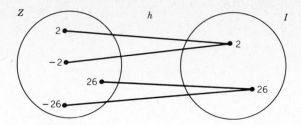

Figure 2.11

The circle on the left represents the domain, the one on the right represents the range, and the arrows show the associations defined by *g*.

Mathematically, there are many functions with which you may be familiar. Consider the function $h:Z \to I$, where Z is the set of integers, I is the set of positive integers, and $h(x) = x$ for every x in Z (where h is the *absolute value function*). For each x in Z, h maps it to a unique number in I. However, it is important to note that for each element in I, there can be more than one element in Z which is mapped to that element by h. For example, both -3 and 3 are mapped by h to 3. What does this mean? In a sense, the uniqueness only goes in one direction. Let's look at the diagram for h in Figure 2.11.

How can we express the uniqueness in the function definition? Since we are dealing with a relation R on a set S, this function obeys the property that, for all a, b, and c in S,

(a, b) in R and (a, c) in R imply that $b = c$

In function notation, this means that if f is a function mapping a set A to a set B, then whenever x is an element of A then $f(x)$ must be a unique element. Thus, there cannot be two elements y and z in B with $f(x) = y$ and $f(x) = z$ but with y not equal to z. Often, this test of a function is written as the contrapositive: if $x = y$ in A, then $f(x) = f(y)$ in B.

TYPES OF FUNCTIONS

As we saw with h above, sometimes more than one element in the domain maps to a single element in the range. If, however, distinct elements of A have distinct images in B whenever $f:A \to B$, then we say that f is a **one—to—one** function from A to B. Thus, if x and y are distinct elements of A, then $f(x) \neq f(y)$. A mapping or function f that is one—to—one is also

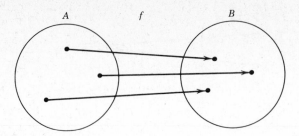

Figure 2.12

referred to as an **injection.** By looking at the diagram of a one—to—one function such as in Figure 2.12, you can see that the arrows showing the relationship are such that there are never two arrows pointing to the same element in the range. (For *any* function, there is never a point in the domain that has two arrows originating from it.)

Suppose *f* is a function mapping a set *S* to a set *Y*. Suppose further that for every element *y* in *Y*, *y* can be expressed as the image of some *x* in *S* under *f*. We describe that by writing $f(S) = Y$, and this means that *Y* is the image of the entire set *S* under the mapping *f*. Relationally, we can describe this property in the following manner. If *R* is a relation on a set *S*, then for each *a* in *S*, we can find an element *b* in *S* such that (a, b) is in *R*. Are all functions like that? Consider the set *Z* of integers. If we define a function $f:Z \rightarrow Z$ by the rule $f(x) = x \times x$ for all *x* in *Z*, then this function *f* does not map into every element of *Z*. Any element of *Z* which is negative, for instance, cannot be represented as the square of an integer and so is not the image of an integer under *f*. However, define $g:N \rightarrow I$ by $g(x) = x + 1$. (Recall that *N* is the set of non—negative integers and *I* the set of positive integers.) Then, *g* indeed expresses every element of *I* as an image of something in *N*. Such a mapping is said to map *N onto I*, whereas *f* only maps *Z into Z*. More formally, a function $f:A \rightarrow B$ is said to be **onto** or a **surjection** if and only if for every element *b* in *B* there is an element *a* in *A* such that $f(a) = b$. Sometimes, the diagrams are drawn as in Figure 2.13 to show the difference between into and onto maps.

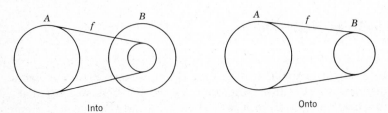

Into   Onto

Figure 2.13

Can a mapping or function be both one—to—one and onto? Certainly. The mapping g above is an example of such a function. A mapping $f: A \rightarrow B$ that is both one—to—one and onto is called a **one—to—one correspondence** or a **bijection.**

EXAMPLES OF FUNCTIONS

Let us examine several examples of functions to see whether they are one—to—one, onto, or bijections. For the first example, let $f:R \rightarrow R$ map the real numbers to the real numbers according to

$$f(x) = \sin x$$

A graph of f is shown in Figure 2.14. What is the domain of this function? Any real number can be used in this definition, so the domain of f is R itself. What about the range? We want to know the set of numbers to which f takes a real number by computing its sine. If you recall, the sine of a number is always between -1 and 1. (If you are not familiar with trigonometric functions, fear not. You can refer to the graph to ascertain the properties of the sine function discussed here.) In fact, for every real number y such that $-1 \leq y \leq 1$, we can find an x so that $\sin x = y$. This means that the range of f is the entire interval from -1 to 1, including the endpoints. We say that f maps R *into* R but *onto* the interval from -1 to 1. Is f a one—to—one function? If it were, then examination of any point in the range would reveal exactly one number in the domain which maps to it. If y is between -1 and 1, is there only one x in R which maps to y? Clearly not. Zero, for instance, is the sine of any multiple of pi, so we have many numbers in the domain going to a single element in the range. Thus, f is not one—to—one.

Let us define another function j:

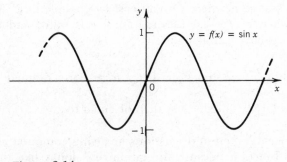

Figure 2.14

$$j(x) = 12\,x + 7$$

It is clear that this function is defined for any real number, so the domain of j is R. Does j map onto the entire set R? If so, then given any y in R, we can find an x in R such that $j(x) = y$. Let us find out what that x is. If y is in R, then we want y to be equal to $12\,x + 7$ for some x. Solving the equation

$$y = 12\,x + 7$$

for x, we find that $x = (y - 7)/12$. Since $j\,[\,(y - 7)/12\,] = y$, this means that every element of R is the image of an element in R under j. Hence, j maps R onto R. Is the mapping j also one–to–one? If so, then distinct elements in the domain would map to distinct elements in the range. Mathematically, this means that $x \neq y$ in the domain R implies that $j(x) \neq j(y)$ in the range. Recall from Chapter 1 that the contrapositive of this statement is equivalent in truth value:

If $j(x) = j(y)$, then $x = y$.

It is in this contrapositive form that we can test to see if j is one–to–one. Assume then that $j(x) = j(y)$. Then, by the definition of j, this means that

$$12\,x + 7 = 12\,y + 7$$

By subtracting 7 from each side,

$$12\,x = 12\,y$$

and so x must be equal to y. Thus, we have shown j to be both one–to–one and onto, so j is a bijection.

Let us examine a nonnumerical example. Let q be a function which maps the set of theatre patrons to the set of seats in a particular theatre for a given performance. Clearly, q must be a one–to–one function, or else we would have a seat somewhere with at least two people in it! Is q an onto function? Only when every seat is sold. Thus, q is onto (and therefore a bijection) when we have a full house.

We examine one final function. Let $k{:}R \rightarrow Z$ be defined as follows:

$k(x) = $ the largest integer less than or equal to x

Such a function occurs often in discussions involving computer science; it is known as a **step** function because the graph of k looks like the graph in

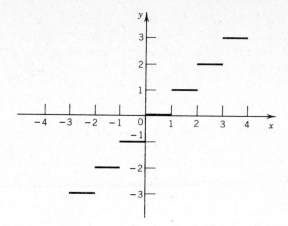

Figure 2.15

Figure 2.15. That the function k is onto Z is easy to see, since for any z in Z, the real number z itself is mapped to z by k. However, it is equally clear that k is not one$-$to$-$one, since the entire interval from $z - 1$ to z (but not equal to $z - 1$) is mapped by k to z when z is an integer.

The exercises at the end of this chapter will give you a lot of practice in deciding whether functions are injections, bijections, or surjections. In addition, the exercises examine special types of functions known as partial functions and total functions.

THE IDENTITY FUNCTION

Recall that the identity relation is a special kind of relation in which an element is related exactly to itself. In a similar manner, we can define an identity function which maps an element in its domain exactly to itself. Thus, the function $I_A: A \rightarrow A$ is the **identity function on** A if and only if $f(x) = x$ for every element x in A. It is clear that the identity function is a bijection from a set to itself. Exercises 20 and 21 will show you why the identity function can be useful.

PRESERVING OPERATIONS: MORPHISMS

Suppose we have a function f defined on a set A. Suppose further that there is a binary operation ∇ defined in A such that if a and b are two elements of A, then $a \nabla b$ is defined to be another element of A. Such an operation is said to be **closed** in A, since the result is still in A. Let us look at how f interacts with this operation ∇. Since a and b are elements of the domain of f, we know that $f(a)$ and $f(b)$ are defined. Moreover, we know that $f(a \nabla b)$ is defined. Suppose the set Y is the range of f, namely, $f(A) = Y$. Then $f(a)$, $f(b)$, and $f(a \nabla b)$ are in Y. If there is a binary operation $\oplus$ defined in Y, then it makes sense to consider $f(a) \oplus f(b)$. There are special mappings which preserve the operation from one set to another, so that whenever $a \nabla b$ is defined in the domain, $f(a) \oplus f(b)$ is defined in the range, and $f(a \nabla b) = f(a) \oplus f(b)$. Such functions are called *homomorphisms*. More formally, if $f\colon A \to Y$ is a function with domain A and range Y, and if ∇ and $\oplus$ are binary operations defined on A and Y, respectively, then f is a **homomorphism from** (A, ∇) **to** $(Y, \oplus)$ if and only if

$$f(a \nabla b) = f(a) \oplus f(b)$$

for every pair of elements a and b in A. What does it mean for a function to be a homomorphism? It means that the operation ∇ in A works the same way that $\oplus$ does in the image of A. Let us look at some examples to get a better intuitive idea. Recall that the function j above was defined as $j\colon R \to R$ where

$$j(x) = 12x + 7$$

The operation "$+$" is a binary operation on real numbers. Let us see if it is preserved under j. If x and y are real numbers, then certainly $x + y$ is in R. We know that

$$j(x) = 12x + 7$$

and

$$j(y) = 12y + 7$$

so we can look at $j(x + y)$. We have

$$\begin{aligned} j(x + y) &= 12(x + y) + 7 \\ &= 12x + 12y + 7 \\ &\neq (12x + 7) + (12y + 7) \end{aligned}$$

so if "$+$" is the operation in both domain and range, we see that the operation is not preserved by j. In other words, j is not a homomorphism from $(R, +)$ to $(R, +)$. Verify for yourself that j is also not a homomorphism from $(R, \times)$ to $(R, \times)$.

Let us try another example. Define the function $p:R \rightarrow R$ by

$$p(x) = 32\,x$$

Then, let us use " + " as our operation and see whether it is preserved. We have $p(x) = 32\,x$, $p(y)(= 32\,y$, and

$$p(x + y) = 32(x + y)$$
$$= (32\,x) + (32\,y)$$
$$= p(x) + p(y)$$

so the operation is indeed preserved. Thus, p is a homomorphism from $(R, +)$ to $(R, +)$. Is this also true for $(R, \times)$ to $(R, \times)$? We see that

$$p(x \times y) = 32(xy)$$
$$\neq (32\,x) \times (32\,y)$$

so p is not a homomorphism from $(R, \times)$ to $(R, \times)$.

The sets and the operations do not have to be identical, as in the examples above. Let A be the set $\{\ 0, 1\ \}$ and B the $\{\ \text{TRUE, FALSE}\ \}$ set. Further, define on A the operation of binary addition below:

+	0	1
0	0	1
1	1	0

In other words, we have addition of binary numbers; however, here we have $1 + 1 = 0$, instead of 10 as with regular binary addition. On the set B, we use as our operation the exclusive or that was introduced in Chapter 1. This operation can be depicted as follows:

XOR	T	F
T	F	T
F	T	F

Finally, define the function $q:A \rightarrow B$ as

$$q(0) = F$$
$$q(1) = T$$

Is q a homomorphism from $(A, +)$ to (B, XOR)? Let us look at the following cases:

1. $q(0 + 0) = q(0) = \text{F}$, but $q(0) \text{ XOR } q(0) = \text{F XOR F} = \text{F}$
2. $q(1 + 1) = q(0) = \text{F}$, but $q(1) \text{ XOR } q(1) = \text{T XOR T} = \text{F}$
3. $q(1 + 0) = q(1) = \text{T}$, but $q(1) \text{ XOR } q(0) = \text{T XOR F} = \text{T}$
4. $q(0 + 1) = q(1) = \text{T}$, but $q(0) \text{ XOR } q(1) = \text{F XOR T} = \text{T}$

In each case, then, q preserves the operation, so q is indeed a homomorphism.

TYPES OF MORPHISMS

What else can we say about q? Is it one−to−one or onto? Yes, and this makes it a special type of homomorphism. A homomorphism $f: (A, \nabla) \rightarrow (B, \oplus)$ is said to be a

monomorphism if the function f is an injection
epimorphism if the function f is a surjection
isomorphism if the function f is a bijection

Also, if $A = B$ and $\nabla = \oplus$, we say that f is an **endomorphism**; a bijective endomorphism is an **automorphism.** The exercises will test your talents in deciding which kind of morphism various functions are.

Why are all of these morphisms important? When they preserve operations, they also preserve some of the properties of one set as the elements are transferred by the function to the other set. This means that defining an appropriate function and proving that it is some sort of morphism can be used as a new proof technique. If we know a lot about a particular set and its operations, and we map that set and its operation onto a new and different set and another operation with a morphism, we can immediately conclude that properties generated in the first set by the first operation are also true in the second set with the second operation. This can not only save us a lot of time and energy in proving things, but it can also give us clues as to the nature of a set with which we are unfamiliar. Two sets may seem completely unrelated, but the existence of a homomorphism means that somehow, disguised differently, they work in the same way. In the chapters that follow, you will see how the clever definition of a function or homomorphism will yield an abundance of information about the sets involved.

EXERCISES

1. Let A be the set { 1, 2, 3 } and B be { ∇, ■ }. Write out all of the pairs in $A \times B$. In general, show that if a set A has m elements and B has n elements, then $A \times B$ has $m \times n$ elements.

2. Finish the proof of Theorem 2.1.

3. In this chapter, we introduced the relation known as congruence modulo n. Consider the set of integers Z and break Z up into three subsets:

$A = \{ x \in Z \mid x$ divided by 3 has remainder 0 $\}$
$B = \{ x \in Z \mid x$ divided by 3 has remainder 1 $\}$
$C = \{ x \in Z \mid x$ divided by 3 has remainder 2 $\}$

Show that { A, B, C } form a 3−partition of Z. Also, show that { A, B, C } is the same as { [1], [2], [3] } for the relation congruence modulo 3.

4. PROGRAMMING PROBLEM: Read in a sequence of up to 20 letters. These letters comprise the elements of the set Q . Next, read in a set of ordered pairs of letters. These pairs define a relation R on $Q \times Q$. Test the relation R to see if it is

reflexive	transitive
irreflexive	an equivalence relation
symmetric	a function
asymmetric	a one−to−one function
antisymmetric	an onto function

5. A relation R relating a set S to itself is called **circular** if $x \, R \, y$ and $y \, R \, z$ imply that $z \, R \, x$ for x, y, z in S. Show that R is reflexive and circular if and only if R is an equivalence relation.

6. If R is a relation such that R is a subset of $A \times B$, then the **converse** of R is the relation $R^{\sim}$, where $R^{\sim}$ is a subset of $B \times A$ and $y \, R^{\sim} \, x$ if and only if $x \, R \, y$. If S is a relation where S is a subset of $B \times C$, then the **composite** of R and S is the relation $R \bullet S$ where $R \bullet S$ is a subset of $A \times C$ described by

$x \, (R \bullet S) \, z$ if and only if for some y in B, $x \, R \, y$ and $y \, S \, z$.

Show that

$R \bullet (S \bullet T) = (R \bullet S) \bullet T$
$(R \bullet S)^{\sim} = S^{\sim} \bullet R^{\sim}$

7. Describe the distinction between an irreflexive relation and an antisymmetric one. Give an example of a relation that is irreflexive but not antisymmetric; of a relation that is antisymmetric but not irreflexive.

8. Let S be a set and let I be the identity relation. If R is a relation which is a subset of $S \times S$, show that

R is reflexive if and only if I is a subset of R.

R is symmetric if and only if $R^\sim$ is a subset of R.

R is transitive if and only if $R \bullet R$ is a subset of R.

9. Let R and T be relations which are subsets of $A \times A$. Show that $R \cup T$, $R \cap T$, and $R^\sim$ are also relations. If R and T are reflexive, is $R \cup T$ reflexive? Is $R \cap T$ reflexive? Is $R^\sim$ reflexive? If R and T are symmetric, are any of $R \cup T$, $R \cap T$ and $R^\sim$ symmetric? If R and T are transitive, are any of $R \cup T$, $R \cap T$ and $R^\sim$ transitive?

10. Let R be a relation which is a subset of $A \times A$. What is the smallest transitive relation T on A with $R \subseteq T$? Why is T called the **ancestral** of R?

11. Prove Theorem 2.3.

12. Test the following relations to see whether they are equivalence relations:

1. $x \, R \, y$ if and only if $|\, x \,| < |\, y \,|$ (for x and y real numbers).
2. If p and q are points of a circle, $p \, R \, q$ if and only if $p = q$ or p and q are diametrically opposed.
3. For x and y real numbers, $x \, R \, y$ if and only if $(x^2) + (y^2) = 1$.

13. Let ■ be a binary operation on a set S which satisfies

$$x \; ■ \; (y \; ■ \; z) = (x \; ■ \; z) ■ \; y$$

Show that ■ is associative and commutative.

14. Define the following sets:

$A = \{\, 2, 4, 6, 8, 10 \,\}$

$B = \{\, 1, 3, 5, 7, 9 \,\}$

$C = \{\, x \in Z \mid x \leq 10 \text{ and } x \geq 0 \,\}$

$D = \{\, x \in Z \mid x \leq 10 \text{ or } x \geq 0 \,\}$

$E = \{\, x \in Z \mid x \leq 10 \,\}$

Find the union of A with each of B, C, D, and E. Find the intersection of A with each of B, C, D, and E. Find $A \cup (C \cap D)$, $B \cap (A \cup E)$, and $(C \cup D) \cap (A \cup E)$.

15. Prove that for any sets A, B, and C,

$$A \cap (B \cup C) = (A \cap B) \cup (A \cap C)$$

This is known as the **distributivity of intersection over union.**

16. Show that the operation of union on sets is associative. Show that the intersection of sets is an associative operation.

17. Finish the proof of Theorem 2.3 to show that De Morgan's Laws hold for union, intersection, and complement of sets.

18. If A and B are subsets of a set S, then we define the **symmetric difference of A and B** to be the set

$$A \, \nabla \, B = (A \cap \bar{B}) \cup (\bar{A} \cap B)$$

Show that the operation ∇ is associative and commutative. Show that intersection is distributive over ∇.

19. We say that a relation R is a **partial function** if R is a subset of $A \times B$ such that whenever (a, b) is in R and (a, c) is in R, then $b = c$. A relation R is a **total function** if R is a subset of $A \times B$ such that for each x in A there is exactly one y in B for which (x, y) is in R. How do partial functions and total functions compare with the definition of function in this chapter? Is there a relationship between a total function and an onto function?

20. Suppose f and g are functions such that $f{:}S \to T$ and $g{:}T \to W$, where S, T, and W are sets. The **composition of f and g** is defined to be the function $g \bullet f$ where

$$g \bullet f{:} \, S \to W \text{ and } g \bullet f(x) = g(f(x)) \text{ for all } x \text{ in } S$$

Show that $f \bullet i_S = i_T \bullet f = f$, and that $(h \bullet g) \bullet f = h \bullet (g \bullet f)$.

21. Show that the following two statements are equivalent:
1. f is a bijection.
2. There is a function g such that $f \bullet g = i_T$ and $g \bullet f = i_S$

22. If f and g are surjections, is $g \bullet f$ a surjection? If f and g are injections, is $g \bullet f$ an injection? If f is a bijection and if $g \bullet f$ is defined, show that

g is an injection if and only if $g \bullet f$ is.

If f is a bijection and if $g \bullet f$ is defined, show that

g is a surjection if and only if $g \bullet f$ is.

23. PROGRAMMING PROBLEM: Let S be the set $\{ a, b, c \}$. Write a program which will define all functions $f{:}S \rightarrow S$. Have your program determine which are injections, surjections, or bijections.

24. For each of the following functions defined on the real numbers, find functions g and h such that $g \bullet f = i$ and $f \bullet h = i$.

1. $f(x) = 7x + 3$
2. $f(x) = 1/x$
3. $f(x) = \cos x$
4. $f(x) = $ the largest integer less than or equal to x

CHAPTER 3

BOOLEAN ALGEBRAS

In the last chapter, we examined sets and operations defined on those sets. Then we defined functions that transformed elements of one set into elements of another set, sometimes preserving operations from the first set to the second. In this chapter, we examine in more detail a set and operations on its elements. You will remember that, given a set, we looked at a binary operation on that set. There is no reason why we can't have several operations defined on the same set at once, with interactions among the several operations carefully described. In fact, we do this all the time with records in a data base. We can add, delete, combine, or change records, and we can consider these to be operations on the records. With the set of real numbers, we have a similar circumstance. We have the operations of addition, multiplication, subtraction, and division all defined on the real numbers, and we know exactly how to evaluate expressions involving some or all of these operations. In the same way, we can consider the collection of all sets, and we operate using union, intersection, and complement according to rules that follow from the definitions of each of those operations. Note in particular that union and intersection are binary operations, while complementation is a unary one; this does not stop us from using all combinations of these operations.

We would like to formalize the concept of a set and the interactions of several operations on it. The entity that results is known as a Boolean algebra, and it is something that underlies a surprisingly large number of applications in discrete structures. In examining particular cases of Boolean algebras, we will see in this and the following two chapters that a Boolean algebra can assume many different forms. The concept of a Boolean algebra, then, draws together what seem like diverse areas to yield some quite powerful results.

BOOLEAN ALGEBRAS

Suppose we have a set S and three operations on it: two binary operations ∇ and $\oplus$, and a unary operation $\sim$. We say that the 4-tuple $< S, \nabla, \oplus, \sim >$ is a **Boolean algebra** if and only if the 4-tuple satisfies all of the following:

1. Associativity: For every a, b, c in S, we have

$$a \triangledown (b \triangledown c) = (a \triangledown b) \triangledown c$$
$$a \oplus (b \oplus c) = (a \oplus b) \oplus c$$

2. Commutativity: For every a and b in S, we have

$$a \triangledown b = b \triangledown a$$
$$a \oplus b = b \oplus a$$

3. Distributivity: For every a, b, c in S, we have

$$a \triangledown (b \oplus c) = (a \triangledown b) \oplus (a \triangledown c)$$
$$a \oplus (b \triangledown c) = (a \oplus b) \triangledown (a \oplus c)$$

4. Idempotency: For every a in S,

$$a \triangledown a = a$$
$$a \oplus a = a$$

5. Absorption: For every a and b in S,

$$a \triangledown (a \oplus b) = a$$
$$a \oplus (a \triangledown b) = a$$

6. Complements: For each a in S,

$$\sim\sim a = a$$

7. De Morgan's Laws: For every a and b in S, we have

$$\sim(a \triangledown b) = (\sim a) \oplus (\sim b)$$
$$\sim(a \oplus b) = (\sim a) \triangledown (\sim b)$$

8. Minimal and maximal elements: There exist elements O and I in S such that for every a in S,

$$O \oplus a = a$$
$$O \triangledown a = O$$
$$I \oplus a = I$$
$$I \triangledown a = a$$
$$a \oplus (\sim a) = I$$
$$a \triangledown (\sim a) = O$$

O is called the **least** or **minimal element**, and I is called the **greatest** or **maximal element**.

It is important to note that some of these properties can be derived from some of the others. Thus, many texts define a Boolean algebra with a minimal set of properties. As an exercise, determine which properties can be derived as a consequence of the others. (See Exercise 1.)

EXAMPLES OF BOOLEAN ALGEBRAS

Let us examine the examples cited above to see whether they are indeed Boolean algebras. First, let R be the set of real numbers. For our unary operation, we will use negation, and for our binary operations, let us use addition and multiplication. Then our candidate 4−tuple is $< R, +, \times, - >$. Can we verify the properties required to make our 4−tuple a Boolean algebra?

1. Associativity:, We know that for any real numbers a, b, and c we have

$$a + (b + c) = (a + b) + c$$

   and

$$a \times (b \times c) = (a \times b) \times c$$

   These properties are clearly true.
2. Commutativity: For any real numbers a and b,

$$a + b = b + a$$
$$a \times b = b \times a$$

   These properties are clearly true.
3. Distributivity: For real numbers a, b, and c,

$$a \times (b + c) = (a \times b) + (a \times c)$$

   This property is clearly true.

However, it is here that we must stop. It is *not* true that addition distributes over multiplication, that is, it is *not* true that

$$a + (b \times c) = (a + b) \times (a + c)$$

because, for example, $10 + (5 \times 2)$ is not equal to $(10 + 5) \times (10 + 2)$. Thus, the 4−tuple $< R, +, \times, - >$ is *not* a Boolean algebra. This is not the only property which is not satisfied. The idempotent property is not satisfied either. Continue through the list of properties of Boolean algebras to see whether any of the other properties are violated.

The next possibility to consider is generated in the following manner. Let A be a nonempty set, and let $P(A)$ be the power set of A. This means that $P(A)$ is the collection of all subsets of A. For our unary operation, we have the complement of a set. For binary operations, we will use union and intersection. Thus, the 4-tuple which is our next candidate as a Boolean algebra is $< P(A), \cup, \cap, ^- >$. From the previous chapter, we know that this 4-tuple satisfies associativity and commutativity. Further, in that chapter we proved De Morgan's Laws. It is easy to show that the complement of a complement of a set is the set itself, since that follows from the definition of the complement. As for idempotents, the definitions of union and intersection tell us that

$$B \cup B = B$$

and

$$B \cap B = B$$

for every subset B of A. Since $B \cap (B \cup C)$ is B and $B \cup (B \cap C)$ is B, absorption is satisfied. Distributivity holds if

$$B \cup (C \cap D) = (B \cup C) \cap (B \cup D)$$

and

$$B \cap (C \cup D) = (B \cap C) \cup (B \cap D)$$

for subsets B, C, and D of A. However, we proved part of this in Chapter 2, and the rest was shown in the exercises at the end of Chapter 2. Thus, all that is left to show is that we have minimal and maximal elements. We know that the power set $P(A)$ has at least A and ϕ as elements, so these are likely candidates for maximal and minimal elements. Since

$$\phi \cup B = B \text{ for every subset } B \text{ of } A$$

and

$$\phi \cap B = \phi$$
$$A \cup B = A$$
$$A \cap B = B$$
$$B \cup \bar{B} = A$$
$$B \cap \bar{B} = \phi$$

then we see that indeed ϕ is the minimal element and A is the maximal element. (What happens if A is the empty set? See Exercise 2.)

PROPERTIES OF THE POWER SET

Thus, the power set of A, in combination with union, intersection, and complement, forms a Boolean algebra. Let us look more closely at the power set. For example, how big is the power set of a set?

THEOREM 3.1 _____

If A is a finite set with n elements, then the power set of A has 2^n elements.

_____

We prove this theorem by induction. To do so, we need to choose the item with which we will step through the cases. Let k be the number of elements of the set A. If k is zero, that is, if A has no elements, then A is the empty set. The power set of A thus has only one element, ϕ, so the number of elements in $P(A)$ is 1. However, $2^0 = 1$, so we have shown our inductive hypothesis true for $k = 0$. Suppose now that for every set S which has t elements, where $t \leq k$, we know that $P(S)$ has 2^t elements. We want to examine a set with $k + 1$ elements and see how many elements are in its power set. Let Q be set with $k + 1$ elements, and let a be one of the elements in Q. Then, the set $Q - \{ a \} = \{ x \in Q \mid x \neq a \}$ has k elements in it. By our inductive hypothesis, then, $P(Q - \{ a \})$ has 2^k elements. Define a new set M where

$$M = \{ E \cup \{ a \} \mid E \in P(Q - \{ a \}) \}$$

The set M contains 2^k elements, all distinct and all distinct from the elements in $P(Q - \{ a \})$. (Why?). It is clear that $P(Q) = M \cup P(Q - \{ a \})$, where $M \cap P(Q - \{ a \}) = \phi$, so we have a 2–partition of $P(Q)$. Thus, the number of elements in $P(Q)$ is equal to the sum of the elements in each of the blocks of the partition, so

number of elements in $P(Q)$ = (number of elements in M) + (number of elements in $P(Q - \{ a \})$

$$= 2^k + 2^k = 2^{k+1}$$

Thus, we are done. ■

_____

It is easy to see from this theorem why some people denote the power set of a set A as 2^A instead of $P(A)$. In Theorem 3.1, we looked at the power set of a finite set. What happens if A is an infinite set?, Since A is always an element of the power set of A and since every element a of A has a corresponding element { a } of P(A), we know that the power set of A also has an infinite number of elements. Does this mean that A and P(A) have the same number of elements? In Chapter 5, we examine the notion of the *size* of a set and look at ways of comparing the numbers of elements in different sets. We will see that A indeed has fewer elements than P(A), even though they are both infinite sets.

BOOLEAN VARIABLES AND FUNCTIONS

In order to apply what we know about Boolean algebras to areas of computer science, we must introduce some related concepts that can be used as tools with which to manipulate the elements of the Boolean algebras. One of these tools is a Boolean or switching function. We say that a variable is a **Boolean** or **switching variable** if the set of possible values it can take is { 0, 1 }. In some places, you may see a Boolean variable defined as one which can be set either to TRUE or to FALSE. These definitions are equivalent. (Why?) Examples of such variables abound. For instance, look at any of the truth tables with which you have worked. Any variable in the truth table is a Boolean variable, because the variable can be either TRUE or FALSE. For any particular property, we can define a variable that is Boolean: let Q be a variable that is 1 if Q has that property and 0 if Q does not. Using Boolean operators with Boolean variables yields other Boolean variables: P AND Q, P OR Q, P NOR Q, and so on. Once we have Boolean variables, we can define a Boolean function on the variables. In the case of two variables, a Boolean function accepts the values of the Boolean variables as input and produces 0 or 1 as output. In other words, a **Boolean function on two variables** maps pairs (x, y) (where x and y are chosen from { 0, 1 }) to the set { 0, 1 }. Any logical operator can thus be viewed as a Boolean function. Instead of writing A AND B, we can consider A and B to be Boolean variables and AND to be a function. Then, the function AND (A, B) is either true (1) or false (0). This is the same as writing $f(A, B) = A$ AND B, which is the functional notation to which you are accustomed. We can write the definition of the function f by enumerating all of the possible results of the mapping as shown in Table 3.1.

This concept can be extended to a function of several variables. Suppose we have n Boolean variables, $x_1, x_2, ..., x_n$. We can define a function on these variables whose result is also a Boolean variable. Such a function would accept a string of n 0s and 1s and produce either a 0 or a 1 as its result. We call this kind of function an n–place Boolean function, and we can describe it rigorously in the following way. Let { 0, 1 }^{n} represent the

A	B	$f(A, B) = A$ AND B
0	0	0
0	1	0
1	0	0
1	1	1

Table 3.1

Cartesian product of n copies of the set $\{ 0, 1 \}$. Then, any element of $\{ 0, 1 \}^n$ is an ordered $n-$tuple $(x_1, x_2, ..., x_n)$, where each x_i is either 0 or 1. A **Boolean function on n variables** is a function which maps $\{ 0, 1 \}^n$ to $\{ 0, 1 \}$. The number of variables n in the $n-$tuple accepted by f is called the number of **places** of f. In the example above, f is a $2-$place Boolean function. You can see that for a $2-$place Boolean function, there are 2^2 rows in the enumeration of the possibilities for f. For a $3-$place Boolean function, there are 2^3, or 8 combinations of values, and, in general, for an $n-$place Boolean function, there are 2^n rows or combinations.

How many different 2–place Boolean functions are there? Each row maps to either 0 or 1, so there are two possibilities for each row. There are four rows, so there are thus $2 \times 2 \times 2 \times 2$ possible different 2–place Boolean functions. Similarly, for n places, there are 2^{2^n} different Boolean functions.

Let us look at all of the $2-$place Boolean functions. We can write them out as Table 3.2 by listing the four possible pairs for A and B and then the 16 different results that can define the Boolean function f. Some of the functions f_i correspond to functions with which we are already familiar. For example, f_1 is just the function that maps A and B to A AND B, so we can denote f_1 simply by AND. Similarly, f_3 maps A and B to the value for A, so f_3 can be denoted by A. Likewise, f_5 is B, f_7 is OR, f_{10} is NOT B, f_{12} is NOT A, f_0 is 0 (because all values are mapped to 0), and f_{15} is 1. In a way, then, each $2-$place Boolean function can be thought of as a *connective between* or an *operator on* Boolean variables. Examine the other functions listed above and see which ones you can recognize as operators. Which one is the XOR? Which one is the NAND?

A	B	f_0	f_1	f_2	f_3	f_4	f_5	f_6	f_7	f_8	f_9	f_{10}	f_{11}	f_{12}	f_{13}	f_{14}	f_{15}
0	0	0	0	0	0	0	0	0	0	1	1	1	1	1	1	1	1
0	1	0	0	0	0	1	1	1	1	0	0	0	0	1	1	1	1
1	0	0	0	1	1	0	0	1	1	0	0	1	1	0	0	1	1
1	1	0	1	0	1	0	1	0	1	0	1	0	1	0	1	0	1

Table 3.2

FUNCTIONAL COMPLETENESS REVISITED

You will recall that in the exercises of the preceding chapter, we talked of sets of operators being *functionally complete*. We can consider the same concept here for Boolean functions. A set S of Boolean functions is **functionally complete** if every Boolean function can be described as a combination of the functions of S. When examining operators on truth values, we saw that the set { AND, OR, NOT } was functionally complete. For the 16 functions f_i enumerated above, is this still true? Exercise 4 will show you that you can indeed express each of the f_i as a combination of AND, OR, and NOT. What about the set { AND, NOT }? If we can show that OR can be expressed as a combination of NOT and AND, then the fact that { AND, OR, NOT } is functionally complete thus implies the functional completeness of { AND, NOT }. We can write A OR B as NOT (NOT A AND NOT B). (Verify this for yourself.)

A TEST FOR FUNCTIONAL COMPLETENESS

The previous examples about functional completeness indicate that it can be quite cumbersome to test for functional completeness by enumerating every case. Is there any other method for determining whether a set of connectives is functionally complete? Let us look at another example. Table 3.3 represents three three—place functions where each F_i is a function of x_1, x_2, and x_3. Do these functions form a functionally complete set? Clearly, trial and error would be very time-consuming, and the more places in the functions, the worse this method would be. Fortunately, there is an algorithm to test a set for functional completeness. It can be a handy tool for logic design and other applications of Boolean functions.

To describe the algorithm, we need to consider the following five classes of functions.

x_1	x_2	x_3	F_1	F_2	F_3
0	0	0	0	1	1
0	0	1	0	1	0
0	1	0	1	1	0
0	1	1	0	0	0
1	0	0	1	1	0
1	0	1	1	0	0
1	1	0	0	0	1
1	1	1	1	1	0

Table 3.3

1. Class 1. Functions closed under true (T).
   This means that if f is such a function,
   $$f(T, T, ..., T) = T \text{ (or equivalently, } f(1, 1, ..., 1) = 1)$$
2. Class 2. Functions closed under false (F).
   This means that if f is such a function,
   $$f(F, F, ..., F) = F \text{ (or equivalently, } f(0, 0, ..., 0) = 0)$$
3. Class 3. Add−type functions.
   A **dummy place** or **dummy variable** in a function is a variable whose value has no effect on the value of the function. For example, if the function f is a function of n variables, and
   $$f(x_1, x_2, x_3, ..., x_i, 0, x_{i+2}, ..., x_n)$$
   $$= f(x_1, x_2, x_3, ..., x_i, x_{i+2}, ..., x_n)$$

   for all possible values of the other variables, then x_i is a dummy variable for f. In the class of add−type of functions, the value of a function depends on whether there is an even or odd number of nondummy places in each row having a T (or a 1). Suppose, for instance, that f is a 5−place Boolean function in this class. Then the value of f is T (or 1) for any row that has exactly one, three, or five T s. In general, then, if f is an add function, $f(x_1, x_2, ..., x_k) = T$ if there is exactly an odd number of T s among $x_1, x_2, ..., x_k$; f is F otherwise.
4. Class 4. Value−inheriting functions.
   The function f is value−inheriting if some column (representing a variable, as in the example above) has two rows, i and j, with the following property:

   Row i of this column contains a T and the function value for this row is T and row j for this column is F and the function value for this row is F.

   For example, let $f(x_1, x_2, x_3)$ be defined as in Table 3.4. Since $f(0, 1, 1) = 0$ and $f(1, 1, 1) = 1$, we see that f is value−inheriting in x_1.

x_1	x_2	x_3	$f(x_1, x_2, x_3)$
0	0	0	1
0	0	1	1
0	1	0	0
0	1	1	0
1	0	0	0
1	0	1	1
1	1	0	0
1	1	1	1

Table 3.4

5. Class 5. Self—dual functions.

In such a function, for each function value column in the truth table, the column reads the same from top to bottom as from bottom to top with each value complemented. How can we test a column of a function to see if it is self—dual? The following procedure will demonstrate such a test, using the example column 01001101.

Take the original column: 01001101

Reverse the order: 10110010

Complement value (each 0 to 1, 1 to 0): 01001101

Since we have arrived back at our original column, the column is self—dual. Similarily, a column reading TTFF or FTFT would be self—dual.

Each of these classes is closed. That means that the composition of any two functions in the same class is again in that class. So, for example, if f and g are in class 2, then $f \bullet g$ and $g \bullet f$ are in class 2. (The proof of this is left as an exercise.)

Having defined the classes, we can now state (without proof) the following theorem.

THEOREM **3.2** _____

POST'S COMPLETENESS THEOREM A set of functions of two—valued logic is functionally complete if and only if

1. at least one of the functions in the set is not in class 1, and
2. at least one of the functions in the set is not in class 2, and
3. at least one of the functions in the set is not in class 3, and
4. at least one of the functions in the set is not in class 4, and
5. at least one of the functions in the set is not in class 5.

_____

Note that some functions are complete in and by themselves. For example, NAND can be used to generate all functions. Likewise, the singleton set { NOR } is functionally complete.

To what classes do some of the common functions belong? Let us look at a few examples:

T_k (a k—place function which is always TRUE or 1): classes 1, 3, 4

F_k (a k—place function which is always FALSE or 0): classes 2, 3, 4

Identity $(I(A) = A)$: classes 1, 2, 3, 4, 5

NOT: classes 3, 5
Equivalence: classes 1, 3
IF—THEN: class 1
AND: classes 1, 2, 4
OR: classes 1, 2, 4

Thus, { IF—THEN, AND, NOT } form a functionally complete set. Note that the number of nondummy places in T_k and F_k is zero. Since zero is even, we say, for example, that T_k is true if the number of nondummy places is even. Also note that the Identity function and NOT are 1—place functions, while Equivalence and IF—THEN are 2—place functions. Must every Boolean function belong to at least one of these classes? Exercise 10 will help you to decide.

APPLICATION OF FUNCTIONAL COMPLETENESS TO LOGIC DESIGN

How does functional completeness fit in with the areas of computer science with which you are familiar? Recall that the arithmetic and logical unit of the computer performs various kinds of operations on binary data. The terms

SSI: small—scale integration
MSI: medium—scale integration
LSI: large—scale integration
VLSI: very large—scale integration

refer to integrated circuits (ICs) built up from what are known as gates. A **gate** is an electrical combination of transistors, resistors, and diodes which represents a Boolean function. **Integration**, then, refers to a combination of gates. Typically, a low voltage (0 volts) represents a logical 0, while a high voltage (5 volts) is treated as a logical 1. A gate will accept several lines as input and produce one line as output. For example, an AND gate has two input lines and one output line, and it operates as shown in Table 3.5. You can see that this translates to the logical values

Input:	A	B	Output:	C
	low	low		low
	low	high		low
	high	low		low
	high	high		high

Table 3.5

Input	Output
high	low
low	high

Table 3.6

which is exactly the truth table for the AND logical operator. Similarly, an inverter, or NOT function, can be represented as Table 3.6.

Small—scale integration is an integrated circuit with only a few gates. Medium—scale integration is more complex, and VLSI is very complex and very powerful. The number of gates in these integrated circuits ranges from several for SSI to several million for VLSI. As with other industries, manufacturers seek to build integrated circuits as economically as possible, subject to constraints on speed, reliability and power consumption. Different kinds of gates vary in their complexity and consumption performance characteristics. For example, AND gates tend to be more complex and require more power than NAND gates. It is desirable to build any kind of circuit from a limited number of types of gates. Functional completeness is thus of great importance in the design of integrated circuits.

EXAMPLES OF LOGIC DESIGN

Earlier in this chapter, we showed that a switching function $f(x_1, x_2, ..., x_n)$ can be expressed in terms of AND, OR, and NOT. Using a (0, 1) notation instead of a (T, F) one, any switching function can be expressed using the symbols + (to represent OR), × (to represent AND), and $^-($ to represent NOT). For example,

$$f(a, b, c) = \bar{a} \times b + c\,(\bar{a} + a \times \bar{c}) + b$$

is a valid way of defining the function f. As in ordinary arithematic, $x \times y$ is often written as xy, and × is performed before +.

Clearly, such functional definitions can get quite complex. A manufacturer of computers may need several thousand realizations of this function in each of the company's computers. How can such a function be realized as economically as possible? The logic design example in Chapter 0 gave you an idea of the magnitude of savings possible. Tools such as Karnaugh maps and the Quine–McCluskey method use the characteristics of switching functions to yield minimal realizations.

GATES AND FUNCTIONAL NOTATION

In order to understand these tools and how they can be applied to this problem, first we must examine more closely the relationship between gates and the switching functions we studied earlier. In particular, let us look at expressions of functions. From Boolean algebra, we know that $x + \bar{x} = 1$. We also know that we can rearrange terms and factors since both $+$ and $\times$ are commutative operations. Consequently, we can use this information to simplify expressions of functions. Suppose f is defined as

$$f(a, b, c, d, e, h) = a\,\bar{b}\,c\,\bar{d}\,\bar{e}\,h + dh\,\bar{b}\,\bar{e}\,ac$$

We compare the two terms to see whether two places differ; that is, we are looking for variables that are complemented in one term of the expression and not in the other. This corresponds to looking for two rows in the defining table which differ only by a single bit. To make this search easier, let us arrange the terms in alphabetic order and then write 1 for a variable that is not negated and 0 for a variable that is. Then, the components of f become

101001

and

101101

Since the fourth digit differs but the rest remain the same, we can factor the original expression in the following way:

$$a\,\bar{b}c\,\bar{d}\,\bar{e}h + a\,\bar{b}cd\,\bar{e}h$$
$$= a\,\bar{b}c\,\bar{e}h\,(\bar{d} + d)$$
$$= a\,\bar{b}c\,\bar{e}h\,(1)$$
$$= a\,\bar{b}c\,\bar{e}h$$

In a similar fashion, recall that

$$a \times \bar{a} = 0$$
$$a + 0 = a$$
$$a + 1 = 1$$
$$a \times 0 = 0$$
$$a \times 1 = a$$

These identities can be used to simplify expressions, too. For example, the expression

$$(a \, \overline{c} d) \times (\overline{a} b)$$

can be reduced to

$$a \, \overline{a} b \, \overline{c} d$$

But, since $a \times \overline{a}$ is 0, the entire expression reduces to 0.

ADDERS

Let us examine in more detail the way in which the logic gate corresponds to the functional notation we have been using. The standard symbols for gates are shown in Figures 3.1 through 3.6. Except for inverters, the number of inputs can vary. Commercially available gates, for instance, typically can be obtained with two, three, four, and eight inputs.

How do we translate from function to gates? Let g be the function

$$g \, (a, b, c) = \overline{a} b + c \, (\overline{a} + a \, \overline{c}) + b$$

The $\overline{a} b$ term is realized with an inverter and an AND gate as pictured in Figure 3.7, and the entire function g is shown in Figure 3.8. This realization is straightforward and easy. There are alternative methods for producing this function, which we will not address in this text.

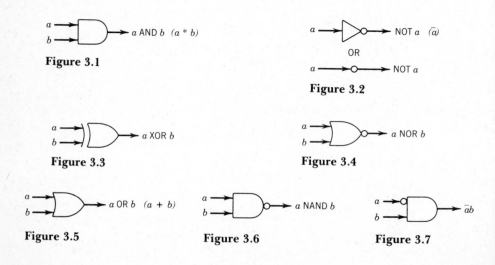

a AND b $(a * b)$

Figure 3.1

NOT a $(\overline{a})$

OR

NOT a

Figure 3.2

a XOR b

Figure 3.3

a NOR b

Figure 3.4

a OR b $(a + b)$

Figure 3.5

a NAND b

Figure 3.6

$\overline{a} b$

Figure 3.7

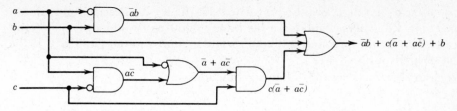

Figure 3.8

Let us consider how addition is performed in a computer. This kind of addition, called **binary addition**, involves the adding together of two binary numbers. For example, to add the two eight−bit numbers here,

00101110
01001011

we add the columns from right to left. Since this is binary addition, if the result of a column addition is greater than 1, we subtract 2 from the column addition, place that number in the column under consideration, and carry a 1 over to the next column on the left. Note that the rightmost column is a special case. There is never any carryover from another column to this column. Thus, in designing a device to perform binary addition on a computer, we can employ two components: one to perform the initial addition on the rightmost column, and another to perform the column additions which must include a carry from a previous column. The first component is called a **half−adder**, since it does not need to incorporate an initial carry. The second component, which *does* incorporate the possibility of an initial carry, is called a **full adder.**

Suppose we are trying to design a half−adder. That is, we want to design a logical device that adds two bits together and produces two outputs: a one−digit sum and a carry digit. If a and b are the two bits we wish to add, we can represent adding them by Table 3.7, where S is the sum of a and b and C is the carry bit.
Figure 3.9 illustrates this function as a circuit. A full adder is more complex, because it must have three inputs: a carry bit from the previous addition, and the two binary digits to be added. As output, the full adder

a	b	S	C
0	0	0	0
0	1	1	0
1	0	1	0
1	1	0	1

Table 3.7

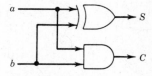

Figure 3.9

produces a sum digit and the bit carried out. Its logical description is shown in Table 3.8, where a and b are the binary digits to be added, C_{in} is the bit carried in by the previous addition, S is the sum resulting, and C_{out} is the bit carried out.

We can write this as the following functions.

$$S(a, b, C_{in}) = \bar{a}\ \bar{b}\ C_{in} + \bar{a}b\ \bar{C}_{in} + a\ \bar{b}\ \bar{C}_{in} + ab\ C_{in}$$

and

$$C_{out}(a, b, C_{in}) = \bar{a}b\ C_{in} + a\ \bar{b}\ C_{in} + ab\ \bar{C}_{in} + ab\ C_{in}$$

How do we generate these functions from the descriptions of the adders? Let us look at S as an example: We examine the function table depicted for the full−adder, and we locate the rows of the table which produce a 1 for S. The first such row has $a = 0$, $b = 0$, and $C_{in} = 1$. We then translate this row into $\bar{a}$ and $\bar{b}$ (negation corresponds to a 0 in the row for that variable's column), followed by C_{in} (corresponding to a 1 in the row for C's column). Similarly, we have rows

0 1 0
1 0 0

and

1 1 1

generating a 1 for S. As before, each 0 translates into a negated term for that column heading, while a 1 translates into just the column heading. Likewise, we can generate a function for C_{out} by examining the C_{out} column of the table, looking at rows which result in a 1 in that column, and translating to terms as above.

a	b	C_{in}	S	C_{out}
0	0	0	0	0
0	0	1	1	0
0	1	0	1	0
0	1	1	0	1
1	0	0	1	0
1	0	1	0	1
1	1	0	0	1
1	1	1	1	1

Table 3.8

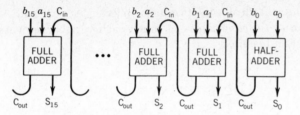

Figure 3.10

Adding two 16—bit numbers would thus require a half—adder for the rightmost addition (since there is no carry into the least significant bit) and then 15 full adders for the other 15 bits, as shown in Figure 3.10. The gate diagram for adding 16 bits is rather complex and is left as Exercise 11. There are better ways to generate a representation, and we will investigate these in the next section.

CANONICAL SUMS AND PRODUCTS

As we have seen previously, switching functions are mathematical ways of expressing Boolean relationships. In the sections above, we have seen how these functions can be combined, and we have also learned a little bit about gates, integration, and logic design. In this section, we will tie this all together to see how switching functions can be realized as gates. In order to show you some very powerful properties of switching functions, we must introduce several new definitions and concepts. Although they represent different ways of talking about the same things we have already discussed, these definitions present a convenient and simple way of thinking about switching functions.

Suppose we have a switching function of n variables, x_1, x_2, ..., x_n. A **literal** is an occurrence of a variable, complemented or uncomplemented. Thus, if f is the function defined by

$$f(x_1, x_2, x_3) = x_1 \overline{x}_2 + x_3$$

then the literals are x_1, $\overline{x}_2$, and x_3. A **minterm** is a product of the n variables, where each variable appears exactly once as a literal. For example, each of the following terms is a minterm:

$$x_1 \overline{x}_2 x_3$$
$$\overline{x}_1 x_2 \overline{x}_3$$

Whereas minterms deal with products, maxterms deal with sums. A **maxterm** is a sum of the n variables in which each variable appears exactly once as a literal. For instance,

$$\overline{x}_1 + x_2 + \overline{x}_3$$

is a maxterm.

We can put these pieces together in the following ways. A **canonical sum of products** or **disjunctive normal form** is an expression represented as a sum of minterms. Similarly, a **canonical product of sums** or **conjunctive normal form** is an expression represented as a product of maxterms. We will see shortly how canonical sums and products can be useful. In general, a **sum of products** (sometimes called an SOP) is a sum of product terms in which no literal appears more than once in any one term. For example, $a\,\overline{b} + \overline{a}c\,\overline{d}$ is an SOP because no literal appears more than once in any term. The same literal may appear in different products; it is only in a single product that we worry about repetition. Likewise, we can look at products of sums. A **product of sums** (POS) is a product of factors where each factor is the sum of literals, and no literal appears more than once in any factor. As with sums of products, a literal may appear in different factors but no literal may appear twice in the same sum. For example,

$$(a + \overline{c})(b + \overline{c} + \overline{d})$$

is a perfectly good POS. Note that a canonical SOP (POS) is also a SOP (POS) but that the reverse is not generally true. For example, $f(x, y) = x + \overline{x}y$ is a SOP but is not a canonical SOP. One function can have many different SOP (POS) representations, but it can have only one *canonical* SOP (POS) representation. It is also important to observe that the canonical representation is a standardized form but may not be the minimal form for a particular function.

Suppose we represent a switching function as a sum of products (SOP). We saw previously that any Boolean functions can be realized as a circuit of gates. Our aim is to minimize the number of gates. Thus, we need to consider the concept of a minimal SOP. A **minimal SOP** is an SOP representing a function such that

1. no other SOP representation of the function has fewer terms, and
2. any other SOP representation of the function that has the same number of terms will not have fewer occurrences of literals.

In a similar way, we define a **minimal POS** to be a POS representing a function; the number of terms is minimal, and no POS representation with the same number of terms has fewer occurrences of literals.

The canonical SOP and POS representations of a function are standardized representations, and they give logic designers an efficient means of realizing Boolean functions as combinations of gates. Thus, the following theorem is quite powerful and very important.

THEOREM **3.3** _____

Every Boolean function of n variables has a canonical SOP and POS representation.

_____

To prove this, let $f(x_1, x_2, x_3, ..., x_n)$ be an n−variable switching function. In tabular form, recall that this means that there are 2^n rows in the table representing the function. To form the canonical SOP, look at all of the rows for which the value of the function is 1. For each of these rows, define a minterm as follows:

If $x_i = 1$ in this row, include x_i in the minterm.
If $x_i = 0$ in this row, include $\bar{x}_i$ in the minterm.

For example, if n is 7 and $f(0, 1, 1, 0, 1, 0, 0) = 1$, then the corresponding minterm is

$$\bar{x}_1 \, x_2 \, x_3 \, \bar{x}_4 \, x_5 \, \bar{x}_6 \, \bar{x}_7$$

This process is illustrated in Figure 3.11. The sum of all these minterms is the canonical SOP. Each minterm has the value 1 for exactly one combination of the values for the x_i s; it will be 0 in all other cases. Further, since the minterms are all summed, the canonical SOP will be 1 if any of the minterms take the value 1, so this canonical SOP represents the function f.

To form the canonical POS, look at all rows for which the value of f is 0. For each such row, we can define a maxterm as follows:

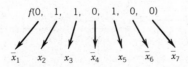

Figure 3.11

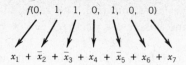

Figure 3.12

if $x_i = 1$ in this row, include $\bar{x}_i$ in the maxterm

If $x_i = 0$ in this row, include x_i in the maxterm.

Again, if n is 7 and $f(0, 1, 1, 0, 1, 0, 0) = 0$, then the corresponding maxterm is

$$x_1 + \bar{x}_2 + \bar{x}_3 + x_4 + \bar{x}_5 + x_6 + x_7$$

Figure 3.12 shows how the term is formed. The product of these maxterms defines the canonical POS. Each maxterm takes the value 0 for one and only one combination of the x_i s; it will be 1 in all other cases. The canonical POS, then, takes the value 0 if any term takes the value 0, so the canonical POS has the value 0 exactly where the rows of the function are 0. ∎

USING CANONICAL REPRESENTATIONS

Now that we are guaranteed our canonical representations for each function, let us look at examples of how they are used in logic design. Suppose we have a function f defined by

$$f(a, b, c) = a\,b\,\bar{c} + a\,\bar{b}\,\bar{c} + \bar{a}b\,\bar{c}$$

In terms of gates, f can be realized as the sequence of gates shown in Figure 3.13. Similarly, a product of sums such as

$$g\,(a, b, c) = (\bar{a} + b + c)(a + \bar{b} + c)$$

can be represented as the sequence shown in Figure 3.14.

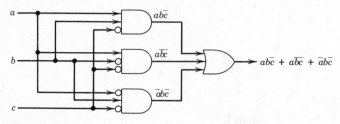

Figure 3.13

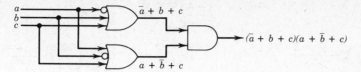

Figure 3.14

These examples are relatively simple. However, when the function has a large number of variables, it may not be possible to obtain a gate with that number of variables. Nevertheless, the functions can still be realized by a series of AND−gates leading into a series of OR−gates (for an SOP), or a series of OR−gates leading to ANDs (for a POS).

THEOREM **3.4** _____

Any SOP or POS representation can be described as a series of AND−gates leading to a series of OR−gates (for an SOP) or as a series of OR−gates leading to a series of AND−gates (for a POS). Further, any such description can be converted to a realization using NANDs instead of ANDs and NORs instead of ORs.

The proof involves constructing the required gates. Suppose we have an SOP representation. We saw above that such a representation can be realized as a sequence of gates. Our proof is an outline of the technique used to transform a function's representation into the required series. Let Figure 3.15 be our initial realization. For each of the lines a, b, c, and so on, Figure 3.16 shows how we can add two inverters without changing the value of the function, since $\bar{x} = x$. This is the same as Figure 3.17. Note that an OR−gate with all inputs inverted is exactly the same as a NAND−gate. We can see this by observing that the gate in Figure 3.18. represents $\bar{x} + \bar{y} + \bar{z}$, which is equivalent to $\overline{(xyz)}$ by De Morgan's Law. Hence, the equivalent is Figure 3.19.

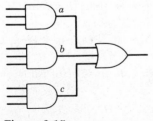

Figure 3.15

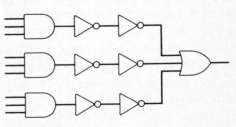

Figure 3.16

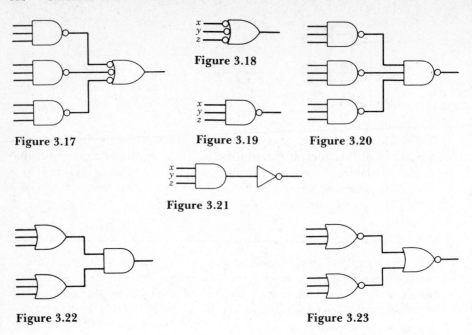

Figure 3.17   Figure 3.18   Figure 3.19   Figure 3.20

Figure 3.21

Figure 3.22   Figure 3.23

Finally, we have Figure 3.20, which contains all NAND−gates. Observe also that Figure 3.21 represents $\overline{(xyz)}$, which is the same as a NAND−gate. The proof for altering a POS is left as Exercise 12. In a similar manner, the gates of Figure 3.22 are converted to gates of the form shown in Figure 3.23, which are NOR−gates. ∎

EXERCISES

1. The beginning of this chapter lists eight properties which define a Boolean algebra. Actually, only a subset of these eight is enough to define the Boolean algebra, because the rest of the properties can be derived from this subset. Determine which properties comprised the "minimal subset" needed for defining a Boolean algebra; then, prove that the remaining properties follow from the minimal subset.

2. Suppose A is the empty set. What is the power set $P(A)$? Is $< P(A), \cup, \cap, ^- >$ still a Boolean algebra? Why or why not?

3. Table 3.9 describes four two−place functions. Is this set of functions functionally complete?

x_1	x_2	g_1	g_2	g_3	g_4
0	0	1	1	1	0
0	1	0	1	0	1
1	0	1	0	1	0
1	1	1	1	0	0

Table 3.9

4. PROGRAMMING PROBLEM: Write a program which will accept as input n, the number of places, and produce as output a table all the $n-$place Boolean functions. Should you place an upper limit on n ? Why?

5. Examine the list of 16 2−place Boolean functions described in this chapter. Rewrite the f_i as combinations of AND, OR, and NOT. For example, f_4 corresponds to IF−THEN. Knowing that IF P THEN Q can be written as NOT P OR Q , how can we write f_4 as a sum or product of the other functions?

6. Do the functions NAND and NOR belong to any of the classes described for use in the Post completeness theorem?

7. For each of the three functions F_1, F_2, and F_3 defined in the example introducing the Post Completeness Theorem, list the classes to which each belongs. Do these three functions form a functionally complete set?

8. Let C_1, C_2, C_3, C_4, and C_5 be the five classes defined in the description of the Post Completeness Theorem. Show that each of these classes is closed under function composition. That is, for each i, show that whenever f and g belong to class C_i, then $f \bullet g$ also belongs to class C_i.

9. Are there any three−place functionally complete functions, that is, three−place functions that, like the two−place NAND and NOR, are functionally complete? If yes, give an example of one. If not, explain why there cannot be any.

10. Give an example of a Boolean function belonging to none of the classes defined in Post's Completeness Theorem.

10. Draw a diagram (using logic gates) of the 16−bit adder as described in this chapter.

12. Finish the proof of Theorem 3.4 by showing that any POS representation can be described as a series of OR−gates leading into a series of AND−gates.

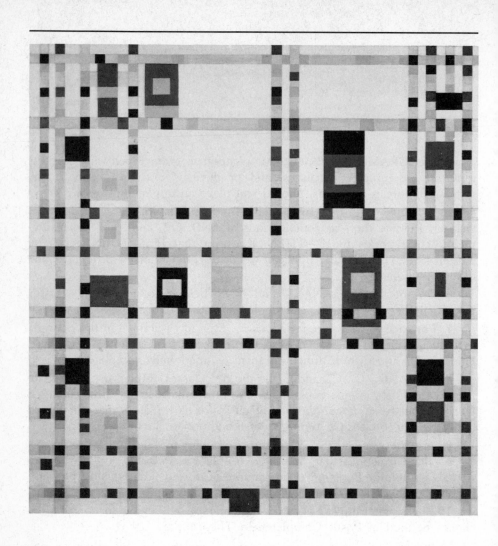

CHAPTER 4

BOOLEAN ALGEBRA
AND LOGIC DESIGN

AND LOGIC DESIGN

In an example in Chapter 0, we saw how minimizing the number of gates in a computer's circuits could lead to a large savings for the computer's manufacturer. Then, in Chapter 3, we looked at a way in which we could translate a gate configuration into a description as a Boolean function. In this chapter, we examine methods for using this functional description to obtain a more efficient logic design.

COVERINGS AND IMPLICANTS

In the previous chapter, we defined what minimal SOPs and POSs are, but we did not investigate how you can tell when a functional expression *is* minimal. For example, the function $f(a, b, c) = ab + \overline{a}c + bc$ reduces to the expression $ab + \overline{a}c$, but it is not clear at this point how we know this.

There are many algorithms and methods for producing minimal SOPs and POSs from given expressions. Here, we investigate two methods. The first, that of generating a Karnaugh map, is a good pencil—and—paper method which can be applied to functions of any number of variables but only works well for functions of up to six variables. For more than six variables, the handling of the variables is done more easily using the Quine–McCluskey method. Quine–McCluskey is applicable to any number of variables and can be implemented on a computer. Although Karnaugh maps allow you to generate both minimal POS and SOP expressions for a function, the Quine–McCluskey method applies more easily to minimal SOPs. Thus, in this chapter we deal primarily with SOPs.

Before we detail the reduction methods, it is useful to explore several concepts related to the expression of a function in terms of its variables. Recall that a function can be defined in a tabular way by listing all of the possibilities for input variables and then adding a final column which indicates whether that particular combination of variable assignments results in a 1 or a 0. For example, a function of three variables can have a tabular definition where there is a column for each of the three variables; then, a fourth column indicates the value of the function for those variable assignments. If f is a function of the variables a, b, and c, then we can represent f as Table 4.1. The asterisks indicate the values of f (a 0 or a 1 in each row). When we rewrote a function such as f as a SOP in Chapter 3, we looked only at the rows of f which were equal to 1. In the example above, if f were equal to 1 for the row where $a = 1$, $b = 1$, and $c = 0$, we would translate that into the term $ab\,\bar{c}$ in the SOP.

We say that two terms in an SOP expression of a function are **adjacent** if the terms are exactly the same except for one variable, and that variable appears complemented in one term and uncomplemented in the other. For example, the terms $ab\,\bar{c}d$ and $abcd$ are adjacent terms. Suppose we have two adjacent terms, x and y, where each term has k variables. Suppose further that the value of f for each of these terms is the same. That is, $f(x) = f(y)$. Since there is only one variable that differs between x and y, and since this variable appears complemented in one and uncomplemented in the other, we can conclude that this variable is redundant. In other words, if x and y differ only in the variable a, then a appears in one and $\bar{a}$ appears in the other. For this combination of variables, then, the value of a makes no difference, since the value of f remains the same in either case. In a sense, then, we can leave out the variable a in the expression of f for x and y. By leaving out a variable, we are on our way toward minimizing the SOP expression of f. Let us look at an example. Suppose $f(a, b, c, d)$ is a function which has as one of the terms in its SOP $ab\,\bar{c}$. What does this say about d? First, we know that since $ab\,\bar{c}$ is a term, then f is 1 whenever a and b are 1 and c is 0. This means that if a is 1, b is 1, c is 0, and d is 0, then f is equal to 1. Likewise, if a is 1, b is 1, c is 0, and d is 1, then f is equal to 1. In other words, the value of d for these settings of a, b, and c does not matter. Furthermore, we can say that the term $ab\,\bar{c}$ represents these two rows of the tabular definition of f. In the same way, we say that in an $n-$place function, a term of $n - 1$ variables represents two rows of the function. We can continue in this vein. A term of $n - 2$ variables represents four rows, a term of $n - 3$ variables represents eight rows, and so on.

a	b	c	f
0	0	0	*
0	0	1	*
0	1	0	*
0	1	1	*
1	0	0	*
1	0	1	*
1	1	0	*
1	1	1	*

Table 4.1

Because a term of fewer variables can represent terms of more variables, we say that the smaller *covers* the larger. One term **covers** another if every 1 represented by the second term in an SOP expression of f is also represented by the first term. Thus, $ab\,\overline{c}$ covers both $ab\,\overline{c}d$ and $ab\,\overline{c}\,\overline{d}$. We can go a step further and limit ourselves to only those terms which yield a 1 or several 1s in the function. An **implicant** is any term which always yields a 1 for the value of f. For example, let

$$f(a, b, c, d) = ab\,\overline{c} + a\,\overline{b}\,\overline{c}d + b\,\overline{d}$$

We know $ab\,\overline{c}$ covers $ab\,\overline{c}d$ and $ab\,\overline{c}\,\overline{d}$, and that f is 1 for $ab\,\overline{c}$. Thus, $ab\,\overline{c}\,d$ and $ab\,\overline{c}\,\overline{d}$ are implicants of f. These are not the only implicants. Note that the implicants do not have to appear in the expression of f. (Exercise 1 will generate the other implicants of this function f.) In the expression of f, note, too, that the term ab is *not* an implicant. It has 1s at places in which f does not. For example, when a, b, c, and d are all set to 1, f is not 1.

A **prime implicant** of a function is an implicant that is not covered by any other implicant. For instance, $ab\,\overline{c}d$ is not a prime implicant because it is covered by $ab\,\overline{c}$. The term $ab\,\overline{c}$ is prime, however. An **essential prime implicant** is a prime implicant that must appear in the expression if the expression is a minimal SOP. As we will see, not all prime implicants are essential.

We will use these concepts in the two methods we consider to generate minimal SOPs. In both cases, the first step of the method is a way of determining the prime implicants of a function by combining adjacent terms. The next step generates the essential prime implicants, and the final step builds the minimal SOP expressions.

Because we will be dealing primarily with implicants, it is important to remember that the larger the set of terms covered by the implicant, the fewer the number of variables in the term comprising the implicant. For f an n−place function of variables x_1, x_2, ..., x_n, we saw above that a term which includes all n variables covers one 1. A term with $n - 1$ variables covers two 1s. In general, a term with $n - k$ variables covers 2^k 1s. Thus, the number of 1s covered by a single term must be a power of 2.

KARNAUGH MAPS

Let us see how we can obtain a minimal SOP or POS of functions of up to six variables by forming a Karnaugh map. A **Karnaugh map** is a grid of

x	y	z	$f(x,y,z)$
0	0	0	1
0	0	1	1
0	1	0	0
0	1	1	0
1	0	0	0
1	0	1	1
1	1	0	0
1	1	1	1

Table 4.2

entries wherein each entry represents one row of the function under consideration. The value of the entry is the value of the function for that row. For example, let us examine a three—place function and its corresponding three–variable Karnaugh map. Define a function f by Table 4.2. The corresponding Karnaugh map looks like Table 4.3. Each combination of x, y, and z is represented by an entry in the Karnaugh map; x is placed on one axis and y and z on another for convenience.

Note that the 11 column lies between the 01 and 10 columns. This is done so that any two adjacent entries differ by one change in one variable. (Two entries are not considered to be adjacent if they are diagonally related. They are adjacent only if they share a horizontal or vertical boundary.) If the columns in the map were to be labeled 00, 01, 10 and then 11, the squares representing (x,y,z) equal to $(0, 0, 1)$ and $(0, 1, 0)$ would be adjacent in the map; however, the representations differ in both the y and z values, that is, there is a difference in two bits. Therefore, by choosing the ordering so that there is only one bit change at a time, we are placing adjacent terms (in the sense defined above) adjacent to one another (in the physical sense).

The function f illustrated above has the canonical representation

$$f(x, y, z) = \bar{x}\,\bar{y}\,\bar{z} + \bar{x}\,\bar{y}z + x\,\bar{y}z + xyz$$

x	yz 00	01	11	10
0	1	1	0	0
1	0	1	1	0

Table 4.3

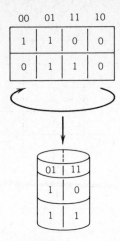

Figure 4.1

These terms represent the implicants of f. The Karnaugh map allows us to pick out adjacent implicants and to determine the prime implicants. Thus, it is easier to examine the Karnaugh map to find the minimal SOP and POS than it is to examine the function itself. First, we note that a single product term with n of the variables eliminated can be represented by a rectangular grouping of 2^n entries. Thus, an implicant will always appear as a rectangular block containing 2^n 1s. If this block is not contained in a larger block whose size is a power of 2, then this implicant is prime (because it is not covered by another implicant). Second, envision the Karnaugh map as being inscribed on a cylinder, slit down one side of the cylinder, and then flattened out, as in Figure 4.1. This makes the map wrap around, in the sense that the leftmost column is really adjacent to the rightmost column. In the same way, consider the top row to be adjacent to the bottom row. Thus, the opposite edges of the Karnaugh map are adjacent. Let us look at some typical Karnaugh maps and see how the terms of the function are represented.

To discover how terms are represented, we must look for common elements. What do we mean by *common elements*? We are trying to find those common elements which describe the location of squares having a value of 1. For example, in Table 4.4, the 1 values are in locations (01, 00), (01, 01), (01, 11), and (01, 10). In all four of these locations, the row labels correspond to the instance where variable w is 0 and variable x is 1. Hence, we say that the common elements are $w = 0$ and $x = 1$, since it is exactly then when we find 1s in the Karnaugh map. Once we have identified the common elements, we convert them into an SOP term for the function represented. A zero indicates that we use the complement of a variable, while a 1 is translated into the variable itself. Thus, $w = 0$ becomes $\bar{w}$, and $x = 1$ becomes x; the term $\bar{w}x$ wx is the result.

wx \ yz	00	01	11	10
00	0	0	0	0
01	1	1	1	1
11	0	0	0	0
10	0	0	0	0

Table 4.4

wx \ yz	00	01	11	10
00	0	0	0	0
01	0	0	1	1
11	0	0	1	1
10	0	0	0	0

Table 4.5

In a similar manner, we see that Table 4.5 has common elements $x = 1$ and $y = 1$, so we form the term xy from the Karnaugh map. Table 4.6 has elements $w = 1$ and $z = 0$, so the resulting term is $w\bar{z}$. Finally, Table 4.7 has as its common elements $x = 0$ and $z = 0$, and we transform this information into the term $\bar{x}\,\bar{z}$.

Let us examine the Karnaugh map concept more closely to see why we are forming terms in this manner. In the map shown in Table 4.8, the rectangle of size 2 that has been circled has common elements $x = 1$ and $z = 1$. The rectangle represents a combining of the terms $x\,\bar{y}z$ and xyz. We know from Boolean algebra that

wx \ yz	00	01	11	10
00	0	0	0	0
01	0	0	0	0
11	1	0	0	1
10	1	0	0	1

Table 4.6

wx $\quad$ yz	00	01	11	10
00	1	0	0	1
01	0	0	0	0
11	0	0	0	0
10	1	0	0	1

Table 4.7

x $\quad$ yz	00	01	11	10
0	0	1	0	0
1	0	1	1	1

Table 4.8

$$x\,\overline{y}z + xyz = xz\,(\overline{y} + y)$$
$$= xz\,(1)$$
$$= xz$$

It is for this reason that the common elements are translated into the term xz.

Table 4.7 looks much more complicated, but it is an example of the very same principle. The common elements are a combining of

$$\overline{w}\;\overline{x}\;\overline{y}\;\overline{z} + \overline{w}\;\overline{x}\;\overline{y}\;\overline{z} + \overline{w}\;\overline{x}y\;\overline{z} + w\;\overline{x}y\;\overline{z}$$

which can be broken down into

$$= (\overline{w} + w)\,\overline{x}\;\overline{y}\;\overline{z} + (\overline{w} + w)\,\overline{x}y\;\overline{z}$$
$$= \overline{x}\;\overline{y}\;\overline{z} + \overline{x}y\;\overline{z}$$
$$= \overline{x}\;\overline{z}\,(\overline{y} + y)$$
$$= \overline{x}\;\overline{z}$$

To obtain a minimal SOP, we look at the 1s in the Karnaugh map and try to cover the 1s in as few rectangles of size 2^n as possible. This means that we try to draw rectangles within the map, where each rectangle contains only 1s. These rectangles are the prime implicants. Remember that the map really wraps around on each end, so a rectangle can include pieces of both sides or the top and bottom. In the example shown in Table 4.9, the four corners make up a perfectly good rectangle.

wx \ yz	00	01	11	10
00	1	0	0	1
01	1	1	0	0
11	0	1	1	0
10	1	1	1	1

Table 4.9

Once we have this covering, we translate each rectangle into a term (as above).

A minimal SOP is a subset of the prime implicants such that all of the 1s of the function are covered by this subset. How do we determine what this subset is? We do this in two steps. Once we have determined the prime implicants, we ask whether any are *essential prime implicants*. For example, is there a 1 in the function which is covered by a single prime implicant? If so, this prime implicant is an **essential prime implicant**. It must be included in any minimal SOP, since no other prime implicant covers this 1.

Let us find the essential prime implicants for the example above. The 1 in the upper right—hand corner of the Karnaugh map (representing $\overline{w}\ \overline{x}y\overline{z}$) is covered by only one prime implicant: the one containing the four corners. Likewise, the term in the lower right—hand center (representing $wxyz$) is also covered by one prime implicant. Thus, these two prime implicants are essential prime implicants: $\overline{x}\ \overline{z}$ (the four corners) and wz (the lower central block of four). However, there are two 1s as yet uncovered. Both are covered by two prime implicants. We try to choose a prime implicant to cover each one as economically as possible. That means that we want to minimize the size of the prime implicant. Thus, we note that we can cover both with a single prime implicant which is a block of size 2. The sum of the terms is exactly the minimal SOP, since it is comprised solely of essential prime implicants. The minimal SOP of this function is thus $f(w, x, y, z) = \overline{x}\ \overline{z} + wz + \overline{w}x\ \overline{y}$.

Sometimes, there are 1s which do not occur in essential prime implicants. In this case, we choose from the prime implicants covering these 1s. Different choices can result in different minimal SOP expressions. Hence, we refer to *a* minimal SOP, rather than *the* minimal SOP. For example, for a Karnaugh map of the form illustrated in Table 4.10, the prime implicants can be represented as Table 4.11.

	yz			
wx	00	01	11	10
00	1	0	0	1
01	1	1	1	0
11	0	1	1	0
10	1	1	1	1

Table 4.10

Table 4.11

Table 4.12

The essential prime implicants are shown in Table 4.12. Each of the remaining 1s is covered by two prime implicants. There are four different choices for picking two prime implicants to cover the remaining 1s, and no choice is preferable to any other. Thus, there are four different minimal SOPs.

KARNAUGH MAPS FOR FIVE OR SIX VARIABLES

This scheme works fine for four or fewer variables, but we need a more manageable representation to handle five— and six—variable functions. In

| | de | | | |
bc	00	01	11	10
00	0	0	0	0
01	0	0	0	0
11	0	1	0	0
10	0	0	0	0

$$a = 0$$

| | de | | | |
bc	00	01	11	10
00	0	0	0	0
01	0	0	0	0
11	0	1	0	0
10	0	0	0	0

$$a = 1$$

Table 4.13

these cases, we use two parts (for five variables) or four parts (for six variables) for each Karnaugh map. In a function of the five variables a, b, c, d, and e, for example, we can draw two Karnaugh maps: one to represent $a = 0$, and another for $a = 1$. Picture the two grids shown in Table 4.13 as situated one atop the other. The two 1 values in this map are adjacent, one above the other when the two grids are stacked. The situation is shown in Figure 4.2.

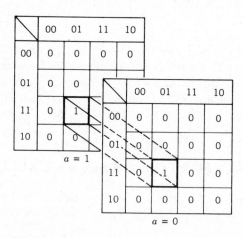

Figure 4.2

For six variables, the representation is more complex. If the variables are a, b, c, d, e, and f, the grid arrangement of the Karnaugh map is made of four components, as in Table 4.14. The four—by—four map for $ab = 01$ is adjacent to the maps for $ab = 00$ and $ab = 11$, but it is not adjacent to the map for $ab = 10$. Picture these as four stacked planes with 00 on top, 01 beneath it, 11 underneath that, and 10 on the bottom.

cd \ ef	00	01	11	10
00	0	0	0	0
01	0	0	0	0
11	0	0	0	0
10	0	0	0	0

$$ab = 00$$

cd \ ef	00	01	11	10
00	0	0	0	0
01	0	0	0	0
11	0	0	0	0
10	0	0	0	0

$$ab = 01$$

cd \ ef	00	01	11	10
00	0	0	0	0
01	0	0	0	0
11	0	1	0	0
10	0	0	0	0

$$ab = 11$$

cd \ ef	00	01	11	10
00	0	0	0	0
01	0	0	0	0
11	0	1	0	0
10	0	0	0	0

Table 4.14

$$ab = 10$$

DON'T—CARES

Sometimes, there are values of a function which are not important. Let us look at an example to see why this kind of function exists. In many instances, we represent a number as its binary equivalent. For instance, Table 4.15 shows the binary equivalents of the first ten nonnegative decimal numbers. The right—hand column is referred to as the **binary coded decimal (BCD)** equivalent of the left—hand column. Note that certain binary combinations (1010, 1011, 1100, 1101, 1110, and 1111) are never used. However, there are other ways of coding the left—hand column which may be more useful. If we let n be our new coding scheme and d be the BCD for the same number, we can devise a code where d is the equivalent of $n + 0011$. This new code, known as **excess—3 code**, looks like the new column of Table 4.16.

Of what use is excess—3 code? It has two major advantages over BCD. First, no combination consists entirely of zeros. Thus, in transmitting data, there is no way for a combination to be misinterpreted as null data. Secondly, excess—3 code is **self—complementing.** This means that the combination for the complement of a digit is the complement of the combination for that digit. This property makes decimal arithmetic easy to implement. Let us see why. For any decimal number k, the **nines complement** of k is defined to be the value $9 - k$. The excess—3 representation of $9 - k$ can be obtained from the excess—3 representation of k by changing all of the 0 bits to 1 and 1 bits to 0. For example, 0 in excess—3 code is represented by 0011, and 9 is represented as 1100. Suppose that you want to subtract two numbers: $a - b$. Using excess—3 representation, we can accomplish this by adding the nines complement of b to a and adding any overflow bit to the result.

Let us work through an example. If $a = 5$ and $b = 1$, then the excess—3 equivalents are

Decimal number	Binary equivalent
0	0000
1	0001
2	0010
3	0011
4	0100
5	0101
6	0110
7	0111
8	1000
9	1001

Table 4.15

Decimal number	BCD	Excess -3
0	0000	0011
1	0001	0100
2	0010	0101
3	0011	0110
4	0100	0111
5	0101	1000
6	0110	1001
7	0111	1010
8	1000	1011
9	1001	1100

Table 4.16

$$5 = 1000$$
$$1 = 0100$$

The nines complement of 1 is 1011, so we perform the following steps:

Begin with 5	1 0 0 0
Add complement of 1	1 0 1 1
Result	1 0 0 1 1
Add overflow bit	* 1
Result	0 1 0 0 $= 4$

In a similar fashion, we can calculate $1 - 5$:

Begin with 1	0 1 0 0
Add complement of 5	0 1 1 1
Result	0 1 0 1 1
Add overflow bit	* 0
Result	1 0 1 1 $= -4$

Note here that 1011 is in 1s complement form. Note, too, that no actual subtraction is performed; the method uses only addition. It is easy to see how much faster such an algorithm is than one which must borrow from the next column to the left. (See Exercise 4.)

Note that the binary entries 1010, 1011, 1100, 1101, 1110, and 1111 do not appear in the preceding table. If we were to extend the BCD column to include these entries, then the corresponding column for excess -3 code would contain six positions in which the code is undefined. Table 4.17 illustrates this. We can think of the table as a tabular definition of a function, and the undefined entries are called **don't$-$cares,** because the function is not defined for those input variables.

BCD	Excess−3		x	y	z	$f(x,y,z)$
0000	0011		0	0	0	0
0001	0100		0	0	1	1
0010	0101		0	1	0	don't-care
0011	0110		0	1	1	1
0100	0111		1	0	0	don't-care
0101	1000		1	0	1	don't-care
0110	1001		1	1	0	0
0111	1010		1	1	1	1
1000	1011					
1001	1100		**Table 4.18**			
1010	don't-care					
1011	don't-care					
1100	don't-care					
1101	don't-care					
1110	don't-care					
1111	don't-care					

Table 4.17

Let us see how don't−care entries in a function definition are handled by the Karnaugh map method. Let f be a don't−care function represented in a tabular way as Table 4.18.

It can be represented by a Karnaugh map as Table 4.19. The question marks indicate the cases in which the result is a don't−care. A square with a question mark can be treated as either a 0 or as a 1, at the user's option. Since we are trying to find a block of four adjacent 1s, we note that the center block would be all 1s if the question mark at (1, 01) were changed to a 1. Table 4.20 shows the center block with this change.

x yz	00	01	11	10
0	0	1	1	?
1	?	?	1	0

Table 4.19

x yz	00	01	11	10
0	0	1	1	?
1	?	1	1	0

Table 4.20

x \ yz	00	01	11	10
0	0	1	1	0
1	0	1	1	0

Table 4.21

To form the equivalent functional representation, we can let two of the squares be 0s and the other a 1 in order to get a minimal SOP. We show this in Table 4.21. It is easy to see here that the minimal SOP representation is z.

In general, this procedure works as follows. First, treat all of the don't−cares as 1s. From this altered Karnaugh map, obtain the set of prime implicants. In this context, an essential prime implicant is a prime implicant that is the only prime implicant covering a 1 (not a don't−care). Thus, once the set of prime implicants is defined, set the don't−care entries back to question marks. Examine the Karnaugh map for 1s covered by a single prime implicant in order to determine the set of essential prime implicants. Finally, choose prime implicants to cover the remaining 1s. The don't−cares need not be covered.

For example, the Karnaugh map of Table 4.22 is considered as in Table 4.23 in order to locate the prime implicants. We have circled the prime

wx \ yz	00	01	11	10
00	1	0	?	0
01	?	?	?	1
11	1	1	0	1
10	?	0	?	0

Table 4.22

wx \ yz	00	01	11	10
00	1	0	1	0
01	1	1	1	1
11	1	1	0	1
10	1	0	1	0

Table 4.23

wx \ yz	00	01	11	10
00	1	0	1	0
01	1	1	1	1
11	1	1	0	1
10	1	0	1	0

Table 4.24

wx \ yz	00	01	11	10
00	1	0	?	0
01	?	?	?	1
11	1	1	0	1
10	?	0	?	0

Table 4.25

wx \ yz	00	01	11	10
00	1	0	?	0
01	?	?	?	1
11	1	1	0	1
10	?	0	?	0

Table 4.26

implicants in Table 4.24. Then, the don't–cares are put back to form Table 4.25. Finally, the essential prime implicants are shown in Table 4.26.

OBTAINING THE POS

To obtain the minimal SOP, we looked for 1s in the Karnaugh map of the function under consideration. This technique takes advantage of the fact that in a Boolean algebra, we always have the relationship

$$x + \overline{x} = 1$$

for any x. Knowing that we were dealing with a sum of products, we could minimize the expression of a function by factoring out sums of this type. To find the minimal POS, we go through similar steps. In this case, since we have a product of factors, each of which is a sum, we want to use the knowledge that $x + \bar{x} = 1$ to help us eliminate some of the factors. Recall that a sum is equal to 0 if and only if every term of the sum is 0. A sum is 1 if and only if any term is a 1. However, in a product, the conditions are reversed. A product is 0 if and only if any term is 0, and the product is 1 if and only if every term is 1. Thus, to minimize the product of sums, we want to examine those values of variables which generate a function value of 0, and we want to invert the conditions.

Recall that the first Karnaugh map we examined represented the function

$$f(x,y,z) = \bar{x}\ \bar{y}\ \bar{z} + \bar{x}\ \bar{y}z + x\ \bar{y}z + xyz$$

This is the minimal SOP. To show its minimal POS, we look again at the Karnaugh map. This is displayed in Table 4.27. We group the 0s in rectangles, just as we had done for the 1s previously. The common elements now (relative to the 0s) are $x = 0$ and $y = 1$ for one of the rectangles, and $x = 1$ and $z = 0$ for the other. The common element $x = 0$ and $y = 1$ generates the term $x + \bar{y}$. This is because the term $x + \bar{y}$ is equal to 0 only for the combination $x = 0$ and $y = 1$; it is equal to 1 for any other combination of x and y. Thus, these common elements yield the terms $(x + \bar{y})$ and $(\bar{x} + z)$, respectively. Consequently, the minimal POS for f is

$$f(x, y, z) = (x + \bar{y})(\bar{x} + z)$$

We can check this (with one step left as an exercise):

$$
\begin{aligned}
f(x, y, z) &= \bar{x}\ \bar{y}\ \bar{z} + \bar{x}\ \bar{y}z + x\ \bar{y}z + \bar{x}yz \\
&= (x + \bar{y} + z)(x + \bar{y} + \bar{z})(\bar{x} + y + z)(\bar{x} + \bar{y} + z) \\
&= ((x + \bar{y})(z + \bar{z}))((\bar{x} + z)(y + \bar{y})) \\
&= (x + \bar{y})(1)(\bar{x} + z)(1) \\
&= (x + \bar{y})(\bar{x} + z)
\end{aligned}
$$

	yz			
x	00	01	11	10
0	1	1	0	0
1	0	1	1	0

Table 4.27

Input				Output				
d_3	d_2	d_1	d_0	e_3	e_2	e_1	e_0	Number represented
0	0	0	0	0	0	1	1	0
0	0	0	1	0	1	0	0	1
0	0	1	0	0	1	0	1	2
0	0	1	1	0	1	1	0	3
0	1	0	0	0	1	1	1	4
0	1	0	1	1	0	0	0	5
0	1	1	0	1	0	0	1	6
0	1	1	1	1	0	1	0	7
1	0	0	0	1	0	1	1	8
1	0	0	1	1	1	0	0	9
1	0	1	0	undefined				10
1	0	1	1	undefined				11
1	1	0	0	undefined				12
1	1	0	1	undefined				13
1	1	1	0	undefined				14
1	1	1	1	undefined				15

Table 4.28

EXAMPLE OF USE OF KARNAUGH MAPS

Let us return to the idea of converting BCD to excess-3 code. To find the minimal SOP, we begin with a tabular function definition and then form the Karnaugh map. Table 4.28 shows the input and corresponding converted output for the decimal numbers 0 through 9. The six four$-$bit input combinations 1010, 1011, 1100, 1101, 1110, and 1111 cannot occur normally, so the operation of a converter function can be defined in whatever way is convenient in order to make its realization in gates as simple as possible. The Karnaugh map corresponding to this function is shown in Table 4.29.

Table 4.29

d_1d_0

d_3d_2	00	01	11	10
00	0	1	1	1
01	1	0	0	0
11	?	?	?	?
10	0	1	?	?

e_2

d_1d_0

d_3d_2	00	01	11	10
00	1	0	1	0
01	1	0	1	0
11	?	?	?	?
10	1	0	?	?

e_1

d_1d_0

d_3d_2	00	01	11	10
00	1	0	0	1
01	1	0	0	1
11	?	?	?	?
10	1	0	?	?

e_0

Finally, the functions resulting which represent this code converter are thus

$$e_3 = d_3 + d_2 d_0 + d_2 d_1$$
$$e_2 = d_2 d_0 + d_2 d_1 + d_2 \overline{d}_1 \overline{d}_0$$
$$e_1 = \overline{d}_1 \overline{d}_0 + d_1 d_0$$
$$e_0 = \overline{d}_0$$

QUINE–MCCLUSKEY METHOD

As mentioned earlier, the Quine–McCluskey method is an alternative to the Karnaugh map for generating a minimal SOP. Whereas a Karnaugh map is awkward to use for more than six variables, the Quine–McCluskey method will work iteratively for any number of variables. In fact, its iterative nature makes the Quine–McCluskey method well suited for implementation on the computer.

Let us work with an example to see how the Quine–McCluskey method operates. Suppose the function f is defined as in Table 4.30. First, we isolate those combinations of variables for which f is equal to 1 or don't–care (a question mark). In our example, there is a 1 at positions 0000, 0010, 0011, 0110, 1011, and 1110, and there are don't–cares at 1000, 1001, and 0111. We want to find adjacencies among these elements, just as we found adjacencies in Karnaugh maps by looking for common elements. To do this, we group these entries according to the number of 1s among the variables w, x, y, and z. For convenience of notation, we consider the string $wxyz$ to be the binary representation of a number, and we note its decimal equivalent in Table 4.31. Because we seek adjacencies, we look for entries in the "binary" column which differ by one bit. These correspond to function definitions which differ exactly in the value of a single variable and are thus adjacent in the functional sense that we defined at the beginning of this chapter. We can save ourselves some work by noting that two entries in the group containing exactly three 1s will never differ by a single bit. (They must differ by

w	x	y	z	$f(w, x, y, z)$
0	0	0	0	1
0	0	0	1	0
0	0	1	0	1
0	0	1	1	1
0	1	0	0	0
0	1	0	1	0
0	1	1	0	1
0	1	1	1	?
1	0	0	0	?
1	0	0	1	?
1	0	1	0	0
1	0	1	1	1
1	1	0	0	0
1	1	0	1	0
1	1	1	0	1
1	1	1	1	0

Table 4.30

Number of 1s	Binary	Decimal
0	0000	0
1	0010	2
1	1000	8
2	0011	3
2	0110	6
2	1001	9
3	1011	11
3	1110	14
3	0111	7

Table 4.31

two or more bits.) Thus, an entry with three 1s need only be compared with entries containing two 1s or four 1s. This step corresponds to finding Karnaugh map blocks of size 2. Since such pairs of entries differ in exactly one variable, we can replace the representation of that variable with an asterisk and drop the variable from the resulting term (just as we did using Karnaugh maps). For example, in f, the 0010 term represents the term $\bar{w}\,\bar{x}y\,\bar{z}$, and the entry 0011 represents $\bar{w}\,\bar{x}yz$. The resultant term, represented as 001*, corresponds to $\bar{w}\,\bar{x}y$, which results from the fact that

$$\bar{w}\,\bar{x}y\,\bar{z} + \bar{w}\,\bar{x}yz = \bar{w}\,\bar{x}y\,(\bar{z} + z)$$
$$= \bar{w}\,\bar{x}y\,(1)$$
$$= \bar{w}\,\bar{x}y.$$

In other words, the term $\bar{w}\,\bar{x}y$ covers both $\bar{w}\,\bar{x}yz$ and $\bar{w}\,\bar{x}y\,\bar{z}$. We continue in a similar fashion to combine pairs of entries, and we generate Table 4.32.

Any entry which could *not* be combined with another corresponds to a 1 in a Karnaugh map which could not be grouped in a block of size 2 or larger.

New term				Combined entries	
0	0	*	0	0,	2
*	0	0	0	0,	8
0	0	1	*	2,	3
0	*	1	0	2,	6
1	0	0	*	8,	9
*	0	1	1	3,	11
*	1	1	0	6,	14
1	0	*	1	9,	11
0	*	1	1	3,	7
0	1	1	*	6,	7

Table 4.32

Term				Combination of	Label
0	*	1	*	2, 3, 6, 7	A
0	0	*	0	0, 2	B
*	0	0	0	0, 8	C
1	0	0	*	8, 9	D
*	0	1	1	3, 11	E
*	1	1	0	6, 14	F
1	0	*	1	9, 11	G

Table 4.33

We can continue this combining process to eliminate another variable. We do this by examining the new terms in the table above and combining those which differ in exactly one variable. At the second and later stages, terms to be combined must have *s at the same locations. This generates an entry of 0 * 1 * which results from combining the 0 0 1 * term (2, 3) and the 0 1 1 * term (6, 7). Since we cannot produce any other terms with variables eliminated (why not?), we list what we have generated as Table 4.33. The terms are assigned labels for convenience. These terms are prime implicants because they are not included in larger combinations.

Next, we list in Table 4.34 all the terms and all the 1s that must be covered by them. (The don't—cares should not be considered here.) An X has been placed in the table when the prime implicant on the vertical axis covers a 1 on the horizontal axis. Since an essential prime implicant is one which must be included in a minimal SOP, we look first for 1s covered by only one prime implicant. For example, 14 is covered only by F. Since F covers both 14 and 6, we delete these entries and reduce our table to Table 4.35.

	0	2	3	6	11	14
A		X	X	X		
B	X	X				
C	X					
D						
E			X		X	
F				X		X
G					X	

Table 4.34

	0	2	3	11
A		X	X	
B	X	X		
C	X			
D				
E			X	X
G				X

Table 4.35

	0	2	3	11
A		X	X	
B	X	X		
E			X	X

Table 4.36

	00	01	11	10	
00	1	0	1	1	wxz
01	0	0	?	1	xyz
11	0	0	0	1	
10	?	?	1	0	

Table 4.37

Next, we search for terms which cover no 1s (such as D) or terms which cover a subset of the 1s covered by some other term (e.g., G covers 11, while E covers 3 and 11; C covers 0, B covers 0 and 2). We eliminate these terms and reduce the table again to Table 4.36. Now only E covers 11 and only B covers 0. Between them, B and E cover all of the remaining 1s. Thus, we include B and E in our minimal SOP, along with F. Hence, the minimal SOP becomes

$$B + E + F = \bar{w}\,\bar{x}\,\bar{z} + \bar{x}yz + xy\,\bar{z}$$

To check this, examine the Karnaugh map in Table 4.37.

At times, the table derived from the Quine–McCluskey method may not yield an obvious result. For example, there are several ways to interpret Table 4.38. In cases such as this one, it is necessary to make an arbitrary choice.

	0	2	3
A	X	X	
B		X	X
C	X		X

Table 4.38

v	w	x	y	z	$g(v, w, x, y, z)$	Decimal
0	0	0	1	0	?	2
0	0	1	0	0	1	4
0	1	0	1	0	1	10
0	1	1	0	0	1	12
0	1	1	0	1	?	13
0	1	1	1	0	1	14
0	1	1	1	1	?	15
1	0	0	1	0	1	18
1	0	1	0	0	?	20
1	1	1	0	0	?	28
1	1	1	0	1	1	29

Table 4.39

Let us follow the course of a second example to get a better feeling for how the Quine–McCluskey method is used. Table 4.39 lists the values of 1 and don't–care for a 5–variable function g. The rightmost column shows the decimal equivalent of the five variables when viewed as a single binary number. As before, we group the variable strings that have the same number of 1s, as in Table 4.40. Next, we look for pairs of entries that differ by exactly one bit, and we create for each such pair an entry that has an asterisk in place of that bit. The result is Table 4.41, where the new terms are grouped by the number of 1s remaining. We repeat this process by grouping the entries in Table 4.41 to eliminate another variable. We place an asterisk in the entry for the variable eliminated, and we form Table 4.42. The terms labeled A through F in Tables 4.41 and 4.42 are prime implicants because they are not included in larger combinations.

Number of 1s	Binary					Decimal
1	0	0	0	1	0	2
	0	0	1	0	0	4
2	0	1	0	1	0	10
	0	1	1	0	0	12
	1	0	0	1	0	18
	1	0	1	0	0	20
3	0	1	1	0	1	13
	0	1	1	1	0	14
	1	1	1	0	0	28
4	0	1	1	1	1	15
	1	1	1	0	1	29

Table 4.40

New term					Combination of	Label
0	*	0	1	0	2, 10	A
*	0	0	1	0	2, 18	B
0	*	1	0	0	4, 12	
*	0	1	0	0	4, 20	
0	1	*	1	0	10, 14	C
0	1	1	0	*	12, 13	
0	1	1	*	0	12, 14	
*	1	1	0	0	12, 28	
1	*	1	0	0	20, 28	
0	1	1	*	1	13, 15	
*	1	1	0	1	13, 29	
0	1	1	1	*	14, 15	
1	1	1	0	*	28, 29	

Table 4.41

New term					Combination of	Label
*	*	1	0	0	4, 12, 20, 28	D
0	1	1	*	*	12, 13, 14, 15	E
*	1	1	0	*	12, 13, 28, 29	F

Table 4.42

From Tables 4.41 and 4.42 we form the prime implicant table shown in Table 4.43; it illustrates those entries that are covered by the prime implicants. As before, no columns are needed for don't—cares. We see from this table that the essential prime implicants are D (the only one that covers 4), B (the only one covering 18), and F (the only entry covering 29). From the remaining labels A, C, and E, we form Table 4.44 of entries not yet covered.

We must pick C for our minimal SOP because it is the only label covering both 12 and 14. Our minimal expression is formed by adding

	4	10	12	14	18	29
A		X				
B					X	
C		X		X		
D	X		X			
E			X	X		
F			X			X

Table 4.43

	10	12	14
A	X		
C	X		X
E		X	X

Table 4.44

$$B + C + D + F$$

which is equal to

$$\overline{w}\ \overline{x}y\ \overline{z} + \overline{v}wx\ \overline{y} + x\ \overline{y}\ \overline{z} + \overline{v}wx + wx\ \overline{y}$$

and we are done.

SEQUENTIAL LOGIC

In the examples generated to illustrate Karnaugh maps and the Quine–McCluskey method, the functions defined were dependent solely on the input received. Given a certain sequence of 0s and 1s, the function defined the output. Such a circumstance is an example of combinational logic. In **combinational logic**, the output depends only on the input values provided; the function processes all input values at once and produces a result in one step. However, there is a different kind of logic that we would like to consider.

Suppose we are traveling from Knoxville, Tennessee, to Sarasota, Florida, by way of Atlanta, Georgia. When we arrive in Atlanta, the connecting flight we choose depends on a number of things. It depends not only on what connecting flights are scheduled, but also on whether our flight from Knoxville arrived in Atlanta on time, whether any of the connecting flights to Sarasota have been delayed or canceled, and whether there is still a seat available on any of the flights that we might take. In this case, the output of our flight–choosing function depends on the historical sequence of input variables. If we arrive late, some flights that we would not normally have considered now become possible choices. We can think of our choice function going from one condition to the next, dependent the values of the historical variables. Such a logical system of making choices is called **sequential logic**. This name is used because the logical choice–making progresses sequentially through the input values.

The way in which we add numbers by hand is another example of sequential logic. First, we sum the rightmost digits, write down a resulting digit, remember a carry digit, and move to the next digits to the left. We

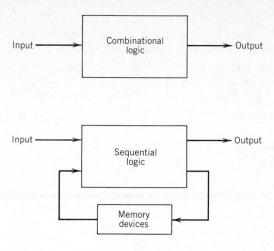

carry with us the history (the carry digit) from the previous part of the addition, and we work sequentially (digit by digit) through the input.

We can view these two logical systems, combinational and sequential, in the following fashion:

The memory devices shown in the sequential logic illustration are often described as "flip—flops." A **flip—flop** is a device that produces one output, and that output must be either a 0 or a 1. You can think of a flip—flop as anything that has exactly two conditions. For instance, a light switch is either on or off, so its two states can represent a flip—flop, as can any other such binary device. Since there are exactly two possibilities represented by one flip-flop, there are 2^n states or possibilities represented by n flip-flops. For example, the states of two flip-flops can be represented as 00, 01, 10, or 11. How can a flip—flop be used in our decision at the Atlanta airport? Each connecting flight's information can be represented by flip—flops. A 0 or 1 can indicate whether or not the flight has left, and a second 0 or 1 can indicate the availability of a seat on the plane. If we arrive in Atlanta at time T, we can check the values of these flip—flops at time T. The memory in sequential logic holds some piece of data from time $T - 1$ to time T. Thus, in sequential logic, the output and the value sent to the memory at time T are functions of the inputs at time T and the memory output at time T. The input to the memory at time T determines the output from the memory at time $T + 1$.

Recall that our examples of logic design using gates dealt with two—bit adders and half—adders. We showed that two—bit adders can be cascaded together to add any number of bits. In this manner, a 16—bit adder would require 16 input lines to enable two 16—bit numbers to be added together. This kind of adder is called a **parallel adder,** because all bits of each input number can be fed to the adder at the same time, that is, in parallel. Suppose, however, we want to build an adder which has only two input lines: one for each input number. The input numbers A and B to be added can be represented as 16 bits each:

$$A = a_{15} \, a_{14} \, a_{13} \, ... \, a_1 \, a_0$$
$$B = b_{15} \, b_{14} \, b_{13} \, ... \, b_1 \, b_0$$

Thus, over the two input lines is passed a sequence of pairs of bits:

$$(a_0, b_0), (a_1, b_1), ..., (a_{13}, b_{13}), (a_{14}, b_{14}), (a_{15}, b_{15})$$

The single output line then carries a sequence of 16 bits to represent the sum of A and B. Presenting the input and output as sequences or series of bits is why we call this type of adder a **serial adder.**

Note that if $A + B$ is represented as

$$c_{15} \, c_{14} \, c_{13} \, ... \, c_1 \, c_0$$

then the value of c_i depends on a_i, b_i and whether there was a carry from the addition of a_{i-1} and b_{i-1}. Thus, at any given time, one of two possible situations can exist:

1. There was no carry from the two previous bits.
2. There was a carry from the two previous bits.

This information, together with the values of a_i and b_i, will tell us what the value of c_i should be and whether there will be a carry into the next two bits.

We can represent the relationship between the two situations by drawing a diagram to depict when and whether one situation changes to the other. Let X represent the first situation (there is no carry) and Y the second (there is a carry). We define a **state** to be a possible situation, and we let a circle represent each state in Figure 4.3. For each state, an arrow is

Figure 4.3

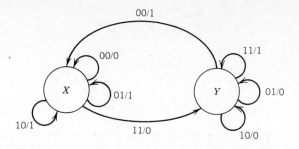

Figure 4.4

drawn for each possible combination of inputs. The state and the inputs determine the destination, and the output is indicated on the arrow. For example, Figure 4.4 means that, when in state X, if $a_i = 0$ and $b_i = 1$, then the output is $c_i = 1$ and we stay in state X. Similarly, if a_i and b_i are both equal to 1, then c_i is equal to 0 but we move to state Y. The arrows are marked with two numbers, a slash, and then another number, where the numbers to the left of the slash indicate the input values a_i and b_i, and the number to the right of the slash is the output value c_i. In this way, the entire 16−bit adder can be represented as Figure 4.5. We begin in state X_0 since there can be no previous carry.

For another illustration of sequential logic, suppose we are defining a parity checker. If we have a sequence of bits, the parity of the sequence is

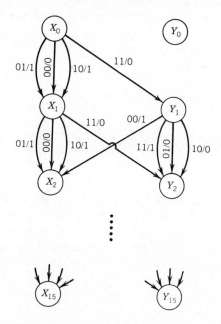

Figure 4.5

b_0	b_1	parity
0	0	1
0	1	0
1	0	0
1	1	1

Table 4.45

determined by counting the number of 1 values in the sequence. If there is an odd number of 1s, we say that this sequence has **odd parity.** If there is an even number of 1s, the sequence has **even parity.** We want to design a three—bit parity checker that will have one input line and one output line. The device will examine three bits in sequence, representing two data bits and a third bit to indicate the parity of the other two. When the two data bits have even parity, the third bit is set to 1, and when the two data bits have odd parity, the third bit is set to 0. Thus, the valid sequences would be those of Table 4.45, while the invalid sequences are

000
110
011
101

These three bits arrive serially, one after another. The first input bit arrives at time T, the second input at time $T+1$ and the parity bit at time $T+2$. Our parity checker is to produce an output of 00 for a valid sequence and an output of 001 for an invalid sequence. Thus, an output of 1 indicates that a bad parity bit has been recognized.

STATE TRANSITION

As with the adders, we can represent the parity checker with a **state transition diagram.** The starting state or situation is represented by the node labeled S.

In Figure 4.6, the notation x/y on an arrow means that the input is x and the output is y. The node labeled S is the situation where nothing in our three—bit sequence has been examined yet. At state A, we have the situation where one bit has been seen, and it was a 0. At state B, one bit has been seen, and it was a 1. For state C, two bits have been seen, and the number of 1—bits was even (either two 1s or two 0s). Finally, for state D, two bits have been seen, and the number of 1—bits was odd.

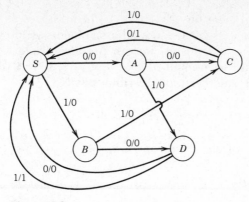

Figure 4.6

If our machine is in state A and the input is a 0, we move to state C and produce a 0 as an output. In state C, if the input is 0, the number of 1–bits is even and the parity bit is 0; this is an incorrect parity bit, so the output is 1. We can describe all of this in a **state transition table.** Such a table (Table 4.46) displays the current state, the inputs to the next state, and the output.

After writing Table 4.46, we want to make a state assignment. That is, we want to assign setting of flip–flops to the states so that each state corresponds to a distinct pattern of flip–flop setting. We have five states. With one flip–flop, we can represent two values, and with two flip–flops, we can represent only four distinct values. Thus, we need three flip–flops to represent our state assignment. One assignment is shown in Table 4.47.

Current state	Input	
	0	1
S	A, 0	B, 0
A	C, 0	D, 0
B	D, 0	C, 0
C	S, 1	S, 0
D	S, 0	S, 1

Table 4.46

State	Flip-flop output		
	y_3	y_2	y_1
S	0	0	0
A	0	0	1
B	0	1	0
C	0	1	1
D	1	0	0

Table 4.47

(Of course, any assignment of y_3, y_2, and y_1 which assures us of uniqueness would suffice, but, as we shall see, not all assignments yield optimal realizations.) From the state assignment, we can produce Table 4.48. From this table, we can see that for an input of $x = 0$, if $y_3 y_2 y_1 = 010$ (state B), we go to state $D = 100$. Another way of writing this is

$$x\, y_3 y_2 y_1 = 0010, \text{ next } y_3 y_2 y_1 = 100$$

If *next* were thought of as a function of x, y_3, y_2, y_1, we could write

$$\text{next } (0, 0, 1, 0) = 1, 0, 0$$

In the same manner, we can rewrite Table 4.48 to look like Table 4.49. The Karnaugh map which corresponds to this table is shown as Table 4.50.

Current state				Next state input 0	1
0	0	0	= S	A=001	B=010
0	0	1	= A	C=011	D=100
0	1	0	= B	D=100	C=011
0	1	1	= C	S=000	S=000
1	0	0	= D	S=000	S=000

Table 4.48

Inputs				Outputs			
x	y_3	y_2	y_1	Y_3	Y_2	Y_1	z
0	0	0	0	0	0	1	0
0	0	0	1	0	1	1	0
0	0	1	0	1	0	0	0
0	0	1	1	0	0	0	1
0	1	0	0	0	0	0	0
0	1	0	1	?	?	?	?
0	1	1	0	?	?	?	?
0	1	1	1	?	?	?	?
1	0	0	0	0	1	0	0
1	0	0	1	1	0	0	0
1	0	1	0	0	1	1	0
1	0	1	1	0	0	0	0
1	1	0	0	0	0	0	1
1	1	0	1	?	?	?	?
1	1	1	0	?	?	?	?
1	1	1	1	?	?	?	?

Table 4.49

xy_3 \ y_2y_1	00	01	11	10
00	0	0	0	(1)
01	0	?	?	(?)
11	0	(?)	?	?
10	0	(1)	0	0

Y_3

xy_3 \ y_2y_1	00	01	11	10
00	1	(1)	0	0
01	0	(?)	?	?
11	0	?	?	?
10	(0	0	0	1)

Y_1

xy_3 \ y_2y_1	00	01	11	10
00	(0	1)	0	0
01	0	?	?	?
11	0	?	?	(?
10	1	0	0	1)

Y_2

xy_3 \ y_2y_1	00	01	11	10
00	0	0	(1)	0
01	0	?	(?)	?
11	(1	?	?	?)
10	0	0	0	0

Z

Table 4.50

We see, then, that for this assignment,

$$Y_3 = x\bar{y}_2 y_1 + \bar{x} y_2 \bar{y}_1$$
$$Y_2 = \bar{x}\,\bar{y}_2 y_1 + x\bar{y}_3 1$$
$$Y_1 = \bar{x}\,\bar{y}_3 \bar{y}_2 + xy_2 y_1$$
$$z = xy_3 + \bar{x}y_2 y_1$$

We have taken a long route to show that our three—bit parity checker can be described by the equations above. However, it is important to note that the state assignments were relatively complex, and the equations show a straightforward way of determining the next state from the existing state.

Let us summarize the steps necessary to transform a state transition diagram into a minimal SOP.

1. First, draw the state transition diagram. Each node will represent a state, and an arrow from one node to another will be labeled x/y if an input of x in the state represented by the first node results in an ouput of y in the state represented by the next node.
2. From the state transition diagram, create a state transition table. List the current states as the first column. There will be a state in this column for each node in the state transition diagram. The second

column, labeled "input," is comprised of a subcolumn for input 0 and another for input 1. The entry in each subcolumn consists of a pair (X, x), where X is the output state given this input, and x is the resulting output (either 0 or 1).

3. The state assignment table represents the states with flip—flop settings. Thus, choose the smallest positive integer k such that 2^k is greater than or equal to the number of states. Using k flip—flops, assign to each state a unique flip—flop setting. Rewrite the state transition table as a table with each state represented as a flip—flop setting. Finally, use this to generate a state assignment table with columns

$$xyk...y_2y_1Y_k...Y_2Y_1z$$

where x is the input, y_k ... y_1 is the flip—flop setting for the starting state, Y_k ... Y_1 is the flip—flop setting for the resultant state, and z is the resultant output. Thus, this table is comprised entirely of 0s and 1s.

4. View the state assignment table as a function definition table. Generate a Karnaugh map or use the Quine—McCluskey method to determine the minimal SOP for this function.

The exercises will give you a chance to try your hand at another state assignment table conversion to a sum of products.

HOW TO FIND AN OPTIMAL REALIZATION

The steps in finding the realization of our sequential machine above are relatively easy to follow. However, there are three major questions related to the search for an optimal realization.

1. First, can we reduce the number of states required? That is, can we construct another machine with fewer states that does the same work?

2. Second, can we split the machine into smaller, simpler machines? If we have a machine that looks like

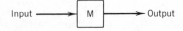

we may be able to decompose the machine into one of two ways:

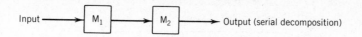

or

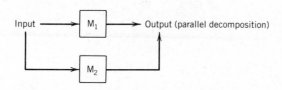

3. What state assignment gives an optimal realization?

Chapter 5 will give us the tools needed to address these questions.

EXERCISES

1. Find all of the implicants of the function represented by the function

$$f(a, b, c, d) = ab\,\overline{c} + a\,\overline{b}\,\overline{c}d + b\,\overline{d}.$$

2. Find the minimal SOP and POS for the Karnaugh maps shown in Table 4.51 and 4.52.

yz	00	01	11	10
wx				
00	1	0	0	1
01	1	1	1	0
11	0	0	1	1
10	1	0	1	1

Table 4.51

yz	00	01	11	10
wx				
00	0	1	1	1
01	1	0	0	1
11	1	0	0	1
10	1	?	0	?

Table 4.52

3. Find a minimal SOP for the following function:

Input			Output
x	y	z	
0	0	0	0
0	1	1	1
1	0	0	1
1	0	1	0
1	1	0	1
1	1	1	1

4. Produce a realization (with minimal POS or SOP) for a two—bit subtractor. The inputs are x_1, x_2, y_1, and y_2, where x_1 and x_2 represent the first two—bit number and y_1 and y_2 represent the second two—bit number. There are to be three output lines: s_1, s_2, and b. The lines s_1 and s_2 represent the two sum bits, and b is a borrow bit. We are subtracting $y_2 y_1$ from $x_2 x_1$. For example, we may have subtraction described by Table 4.53. The borrow bit indicates that a borrow was needed and that the result can be regarded as negative.

5. In the section on state transition, we assigned each of five states to a representation by three flip–flops, y_1, y_2, and y_3. From this table, we generated SOP equations Y_1, Y_2, Y_3, and Z. In a similar fashion, generate different SOP functions for the state assignment table shown in Table 4.54.

6. In this chapter's example of how to obtain the minimal POS, we asserted that

x_2	x_1		y_2	y_1		s_2	s_1		b
0	1		0	0		0	0		0
0	0		0	1		1	1		1
1	0		0	1		0	1		0

Table 4.53

State	y_3	y_2	y_1
S	0	0	1
A	0	1	0
B	1	0	0
C	1	0	1
D	1	1	1

Table 4.54

$$f(x, y, z) = \overline{x}\ \overline{y}\ \overline{z} + \overline{x}\ \overline{y}z + x\ \overline{y}z + xyz$$
$$= (x + \overline{y} + z)(x + \overline{y} + \overline{z})(\overline{x} + y + z)(\overline{x} + \overline{y} + z)$$

Show that this is true.

CHAPTER 5

LATTICES AND THEIR APPLICATIONS

In this chapter, we return to some basic concepts involving sets. These concepts will allow us to decompose and optimize the machine realizations generated by the techniques presented in Chapter 4.

ORDERING SETS

In looking at sets and their elements, we have not paid much attention to ways in which to order these sets. However, it is clear that many sets allow their elements to fit into some sort of ordering scheme. The integers and the real numbers, for instance, are ordered; we can always tell when one number is bigger, smaller, or equal to another number. We have a similar situation with subsets of sets. We can say that one set is a subset of another, and we can compare two sets to see whether one is a subset of the other or the two are equal. Unlike numbers, we cannot always compare two sets and have one be related to the other. For instance, if A is $\{\ 1, 2, 3, 4\ \}$ and B is $\{\ 4, 5\ \}$, neither $A \subseteq B$ nor $B \subseteq A$.

Let us look at the notion of ordering in a more mathematical way. We call a relation R on $A \times A$ a **partial order of A** if it is reflexive, transitive, and antisymmetric. This means that we have

1. Reflexivity: aRa for every element a in A.
2. Transitivity: If aRb and bRc for three elements a, b, and c in A, then aRc.
3. Antisymmetry: If aRb for two elements a and b in A, then we do *not* have bRa unless $b = a$.

It is important to recognize that not all elements of A need be related. In the example cited above, for instance, "is a subset of" is reflexive, transitive, and antisymmetric, but $\{\ 1, 2, 3, 4\ \}$ and $\{\ 4, 5\ \}$ are not related by "is a subset of." The relation "is a subset of" is a partial order nevertheless. If, as with "is greater than" on the set of integers, we have a partial ordering, and

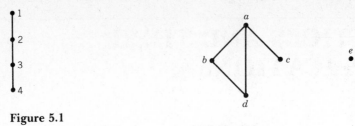

Figure 5.1

Figure 5.2

we can also compare each pair of elements of the set, then we call this a *total ordering*. More precisely, a **partial ordering R on A × A is total** if for every two elements *a*, *b* of *A*, either (*a*, *b*) is in *R* or (*b*, *a*) is in *R*. Sometimes, a total ordering is called a **linear ordering** because we can line up the elements of *A* according to their place in the ordering.

We say that a pair (*A*, *R*) is a **partially ordered set** (or **poset**) if *R* is a subset of *A* × *A* which is a partial order on *A*. It is convenient to represent partially ordered sets using a diagram called a **Hasse diagram.** In such a diagram, a point represents each element of the set *A*. An element *a* of *A* is placed above element *b* and a line is drawn between them if (*a,b*) is in *R*. For example, let *A* be the set of integers { 1, 2, 3, 4 }, and let the relation *R* be defined as *aRb* if and only if $a \leq b$. Then, Figure 5.1 shows the Hasse diagram for this partially ordered set.

Let's look at another example. Let *A* be the set { *a*, *b*, *c*, *d*, *e* }. The relation *R* is { (*a, a*), (*b, b*), (*c, c*), (*d, d*), (*e, e*), (*a, b*), (*a, c*), (*b, d*),(*a, d*) }. (Exercise 7 will convince you that this is a partial order.) We can draw the Hasse diagram for this partially ordered set as Figure 5.2.

We can simplify this somewhat. Because all partial orders are transitive, the fact that *a* is connected to *b* and that *b* is connected to *d* implies that *a* is related to *d*. Thus, we can leave out the connection from *a* to *d* as shown in Figure 5.3. A transitivity appears as a triangle in a Hasse diagram, since we have *aRb, bRc,* and *aRc*. Thus, whenever we see a triangle, we can delete the leg from the highest to the lowest point of the triangle.

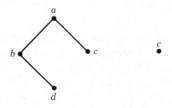

Figure 5.3

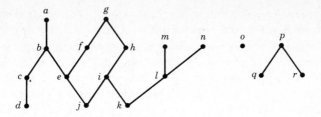

Figure 5.4

Just as we can go from a partial order to a Hasse diagram, we can also use a Hasse diagram to define a partial order. For example, consider Figure 5.4. For each connection of points, we can put that pair of points into our relation. For example, we see that f is above e and that f is connected to e. This means that (f, e) is in R. Is this all we need to do? No. In order to have reflexivity, we must put in a pair (x, x) for every x in the diagram. Further, to have transitivity, we must include all pairs that are implied by intermediate connections. For instance, we have (m, l) and (l, k) in our relation, so we must include (m, k), too. You will have a chance to complete the enumeration of pairs of this partial order in Exercise 8.

UPPER AND LOWER BOUNDS

Let us look more closely at some of the relationships among the elements of a partially ordered set. Let A be a set and R a partial order relation on $A \times A$. If a, b, and c are elements of A, c is said to be an **upper bound** of a and b if (a, c) is in R and (b, c) is in R. Similarly, c is a **lower bound** for a and b if both (c, a) and (c, b) are in R. Elements in a partially ordered set can have many upper bounds and lower bounds.

For example, look at the partial order defined by the Hasse diagram in Figure 5.5. You can see that both a and b are upper bounds for d, that a is an upper bound for both b and c, that both b and d are lower bounds for a, and that e has no upper or lower bound. The element a has a special property with respect to d. Not only is a an upper bound for d but it is an

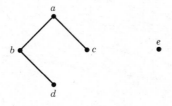

Figure 5.5

upper bound of every other upper bound of d. For a set A and a partial order relation R on A, the **supremum** of A (or the **least upper bound for** A) is that element z in A such that z is an upper bound of elements in A; and, for every x in A, if x is an upper bound, (z, x) is in R. Likewise, the **infimum of** A (or the **greatest lower bound of** A) is that element y of A such that y is a lower bound of elements of A; and, for every x in A, if x is a lower bound, (x, y) is in R. Does every partially ordered set have an infimum and a supremum? If it does, is the infimum unique? Is the supremum unique?

Let us look at an example. Let S be the set of positive integers greater than 1, and let R be the relation defined by xRy if and only if x is a divisor of y. Verify that R is indeed reflexive, transitive, and antisymmetric and thus places a partial order on S. Note, too, that not every pair of elements of S are related by R. For instance, 12 is not a divisor of 17, so they are not related by R. What does the Hasse diagram for this partially ordered set look like? We can see from the diagram in Figure 5.6 that all of the prime numbers are lower bounds, but there is no element of S that forms a greatest lower bound. Thus, this partially ordered set has no infimum. Has it a supremum? (See Exercise 21.)

In many instances, you will see the greatest lower bound of two elements referred to as the **meet** of those elements, and it will be denoted glb(a, b) or $a \wedge b$. The least upper bound of a and b is also called the **join**

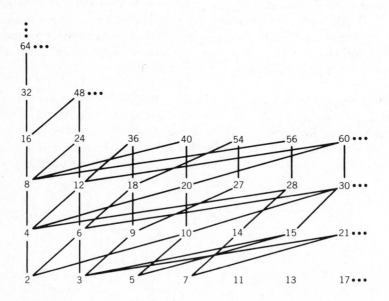

Figure 5.6

of a and b, and it is written as lub (a, b) or $a \lor b$. If an element z is a greatest lower bound of a and b, then we know two things:

1. (z, a) and (z, b) are in R.
2. If (w, a) and (w, b) are in R, then (w, z) is in R.

Similarly, if y is a least upper bound for a and b, then

1. (a, y) and (b, y) are in R.
2. If (a, w) and (b, w) are in R, then (y, w) is in R.

LATTICES

A **lattice** is a partially ordered set in which any two elements a and b have a least upper bound and a greatest lower bound. The partially ordered set represented by the Hasse diagram (a) in Figure 5.7 is not a lattice because a and b have no least upper bound; the partially ordered set in (b) is a lattice.

Consider the set I of positive integers. Define the relation R on I so that (x, y) is in R if and only if x divides y. Then, the join (or greatest lower bound) of x and y is the greatest common divisor of x and y. The meet (or least upper bound) of x and y is the least common multiple of x and y. Is this set a lattice? Yes, because we can always find the least common multiple and greatest common divisor of any two integers.

Often, reference is made to "the" least upper bound or greatest lower bound. This is because, if one exists, it must be unique. We show this in the following theorem.

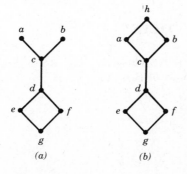

Figure 5.7

THEOREM 5.1 _____

Let A be a set and R a relation which is a partial ordering on A. Let x and y be elements of A. If z is the least upper bound of x and y, then z is unique.

_____

Suppose that z_1 and z_2 are two different least upper bounds for x and y. Then, we know that

$$(x, z_1) \text{ and } (y, z_1) \text{ are in } R$$

because z_1 is an upper bound of x and y. Also,

$$(x, z_2) \text{ and } (y, z_2) \text{ are in } R$$

because z_2 is an upper bound of x and y. Since z_1 is a least upper bound, we know that (z_1, z_2) is in R. Likewise, z_2 is a least upper bound, so (z_2, z_1) is in R. However, this means that both (z_1, z_2) and (z_2, z_1) are in R, which is only possible if $z_1 = z_2$, because R is antisymmetric. ∎

_____

A similar proof (which is left as an exercise) shows the following.

THEOREM 5.2 _____

Let A be a set and R a relation which is a partial order on A. Let x and y be elements of A. If z is the greatest lower bound of x and y, then z is unique.

_____

PROPERTIES OF LATTICES

Let us examine lattices more closely. First, note that the greatest lower bound or least upper bound of two elements a and b could be one of those elements. For example, if a and b are related as in Figure 5.8, then

$$\text{lub}(a, b) = a \vee b = a$$

and

$$\text{glb}(a,b) = a \wedge b = b$$

What about some of the laws regarding operations? If we regard least upper bound and greatest lower bound as binary operations (i.e., considering join and meet as binary operations), what can we say about the properties of the operations? It is easy to verify the following from the definitions of join and meet:

1. Commutativity:
$$a \vee b = b \vee a$$
$$a \wedge b = b \wedge a$$
2. Associativity:
$$a \vee (b \vee c) = (a \vee b) \vee c$$
$$a \wedge (b \wedge c) = (a \wedge b) \wedge c$$
3. Idempotents:
$$a \vee a = a$$
$$a \wedge a = a$$

DUALITY

It is evident that there is some relationship between meet and join. The properties they satisfy are the same. Let us see why. If S is a partially ordered set with order relation R, we can write that partial order as $a < b$ for each pair which is comparable under R. In other words, whenever (a, b) is in R, then we can abbreviate that as $a < b$. Indeed, we can consider $(S, <)$ to be the poset. For each such poset $(S, <)$, we can define another poset $(S, >)$ where $>$ is an abbreviation for the relation R' defined as

$$(a, b) \text{ is in } R' \text{ whenever } (b, a) \text{ is in } R.$$

The relation R' is said to be the **dual** of R, and we say that the poset $(S, >)$ is the **dual** of the poset $(S, <)$. If we have a Hasse diagram for a relation R, then the Hasse diagram for R' is formed by inverting the diagram for R. Figure 5.9 shows an example of this. Suppose A is the set $\{ a, b, c \}$. If the relation R is defined as

Figure 5.8

R

R'

Figure 5.9

$$R = \{ (a, a), (a, b), (c, b) \}$$

then its dual R' is

$$R' = \{ (a, a), (b, a), (b, c) \}$$

If $(S, <)$ forms a lattice, then $(S, >)$ is a lattice, too. How closely related are these two lattices? So closely that we can prove the following theorem:

THEOREM 5.3 _____

> If $(S, <)$ is a lattice and $(S, >)$ its dual, then the join of $(S, <)$ is the meet of $(S, >)$.

To prove this, let a and b be two elements of S, and let z be the join of a and b under the ordering $<$. Then, from the definition of join we know that $a < z$ and $b < z$. However, since $>$ is the dual relation of $<$, then we also know that $z > a$ and $z > b$. Also, if there is a y in S such that $a < y$ and $b < y$, then it must be the case that $z < y$. By duality, this means that for any y in S such that $y > a$ and $y > b$, then $y > z$. But this is just the definition of meet for $>$. ■

Likewise, the next theorem follows from the definitions of meet and join:

THEOREM 5.4 _____

> If $(S, <)$ is a lattice and $(S, >)$ is its dual, then the meet of $(S, <)$ is the join of $(S, >)$.

Of what significance are these theorems? They tell us that whenever we can exhibit a property of meet in one lattice, we automatically know that the property holds for the join of the dual lattice. Also, whenever we know something about the join of a lattice, we know the same thing about the meet of the dual lattice. Clearly, this cuts our work in half. Another reason that these theorems are important is because there are many examples of dual systems in the "real world," some of which you may have encountered. For instance, in electrical engineering, voltage and current, inductance and capacitance, and resistance and conductance are dual pairs. In graph theory, a cycle of a graph corresponds to its dual cut−set. What do the duals of other lattices look like? In the exercises, you will have a chance to examine some of the lattices we have defined and generate their duals.

Let us use the principle of duality in exhibiting other lattice properties. For example, we can show the following:

THEOREM 5.5 _____

For every a in the poset $(S, <)$, $a \vee a = a$ and $a \wedge a = a$.

_____

By definition, we know that $a < a \vee a$. Since $a < a$, this means that a must be equal to $a \vee a$. By the principle of duality, we also know that $a \wedge a = a$. ■

_____

As you recall from Boolean algebras, this is known as the idempotent property. What other properties of Boolean algebras are also properties of lattices? Does a lattice have a unique maximal element or a unique minimal element? The answer is no if the lattice is infinite, that is, if it is defined on an infinite number of points. However, suppose we restrict our discussion to finite lattices. Then, we can show that indeed there is a unique maximal element, and duality will guarantee that we also have a unique minimal element.

THEOREM 5.6 _____

Let $(S, <)$ be a partially ordered set, where S is a finite set. Then there is a unique maximal element.

_____

Suppose S has two maximal elements Y and Z. Then, for any x in S, $x < Y$ and $x < Z$. However, since both Y and Z are elements of S, this means that $Y < Z$ and $Z < Y$, so it must be the case (why?) that Y and Z are equal. ■

_____

Let us denote the maximal element by 1 and the minimal element by 0.

LATTICES AS BOOLEAN ALGEBRAS

So far, the properties we have discussed for lattices are also properties of Boolean algebras. Is a lattice in fact a Boolean algebra? We have not yet

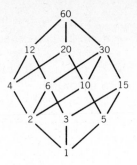

Figure 5.10

satisfied the properties of complementivity and distributivity. It turns out that not every lattice satisfies these properties, so we consider these special kinds of lattices separately. For example, look at the finite lattice of integers illustrated in Figure 5.10. The relationship between elements is that of divisibility: a and b are related if and only if a divides b. This lattice is not a Boolean algebra. Can you tell why not?

COMPLEMENTIVE LATTICES

A **complementive lattice** is one in which for every element x there exists a unique y such that

$$x \wedge y = 0 \text{ and } x \vee y = 1$$

The **complement** of an element x is thus the unique element y such that $x \wedge y = 0$ and also $x \vee y = 1$.

Which of the lattices in Figure 5.11 are complementive? The 0 and 1 can be thought of as special kinds of n−place functions. For any n variables x_1, x_2, ..., x_n, we have a function which is always 0 and a function which is always 1. All other functions fall somewhere between, with function values sometimes 0 and sometimes 1. Hasse diagram A represents a lattice which is not complementive, because b has no complement. However, the lattice represented by B is complementive. Let us see why. Element a has complement d and vice versa. Element b has complement c, since

$$b \vee c = a = 1$$

and

$$b \wedge c = d = 0$$

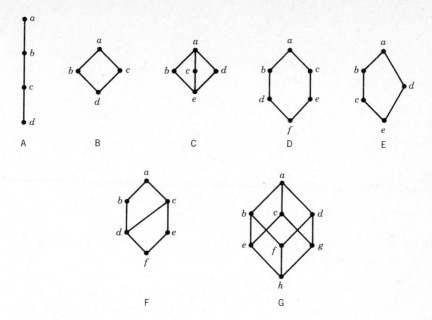

Figure 5.11

The lattice of diagram C has complements that are not unique. In particular, b has both c and d as complements:

$$b \lor c = b \lor d = a$$

and

$$b \land c = b \land d = e$$

Thus, even though every element in lattice C has a complement, C is not a complementive lattice. D also has nonunique complements:

$$b \lor c = b \lor e = a$$

and

$$b \land c = b \land e = f$$

Hence, D cannot be complementive. Likewise, E has nonunique complements for d and therefore cannot be complementive.

Lattice F is noncomplementive for a different reason: c has no complement.

$$c \lor b = a$$

but

$$c \land b = d$$

Also,

$$c \lor d = c$$

and

$$c \land d = d$$

Finally, a careful consideration of joins and meets in lattice G reveals that G is indeed complementive.

As we have seen before with other kinds of complements, the complement of an element is often denoted by that element with a bar over it. Thus, if x is an element, then $\bar{x}$ denotes its complement.

DISTRIBUTIVE LATTICES

A **distributive lattice** S is one in which, for every a, b, and c in S, the distributive property holds:

$$a \land (b \lor c) = (a \land b) \lor (a \land c)$$
$$a \lor (b \land c) = (a \lor b) \land (a \lor c)$$

Which of the lattices in Figure 5.12 are distributive? Distributivity (or the lack of it) is much more difficult to demonstrate than complementivity. A and B are both distributive. For example, to show that B is distributive, we test each case:

1. $a \land (b \lor c) = a \land a = a$

   and

   $(a \land b) \lor (a \land c) = b \lor c = a$

2. $a \land (b \lor d) = a \land b = b$
   and

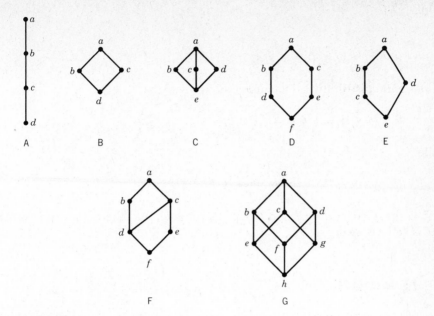

Figure 5.12

$$(a \wedge b) \vee (a \wedge d) = b \vee d = b$$

3. $b \wedge (a \vee d) = b \wedge a = b$

and

$$(b \wedge a) \vee (b \wedge d) = b \vee b = b$$

4. $b \vee (a \wedge c) = b \vee c = a$

and

$$(b \vee a) \wedge (b \vee c) = a \wedge a = a$$

and so on. In lattice C, we have

$$b \wedge (c \vee d) = b \wedge a = b$$

but

$$(b \wedge c) \vee (b \wedge d) = e \vee e = e$$

so C cannot be distributive. Likewise, in D we have

$$b \wedge (c \vee d) = b \wedge a = b$$

but

$$(b \wedge c) \vee (b \wedge d) = a \vee d = d$$

so D is not distributive. Examining E, we see that

$$b \wedge (d \vee c) = b \wedge a = b$$

while

$$(b \wedge d) \vee (b \wedge c) = e \vee c = c$$

so E is not distributive. F and G are distributive, and the demonstration of this is left to Exercise 12.

STATE REDUCTION

A knowledge of lattices will help us to answer the three questions we had posed at the end of Chapter 4. In Chapters 3 and 4, we looked at state diagrams and their realizations as sums of products. Now we want to know if and how we can reduce the number of states in a state diagram. Indeed we can. First, let us see when two states can be considered to be the same. We say that two states A and B in a state transition table or state diagram are **distinguishable** if either

1. for some input x, A and B produce different outputs, or
2. for some sequence of inputs $x_1 x_2 \ldots x_n$, if we start from states A and B, respectively, we get a different sequence of outputs; that is, we get $a_1 a_2 a_3 \ldots a_n$ by starting with A, $b_1 b_2 b_3 \ldots b_n$ by starting with B, and for at least one i where $1 \leq i \leq n$, we have $a_i \neq b_i$.

Two states A and B are **indistinguishable** if they are not distinguishable. This means that if A and B are indistinguishable, we get identical sequences of outputs if we start in either state A or state B.

Let us examine a state diagram to see what we mean by distinguishable and indistinguishable. In the example described by Table 5.1, an input of 1 for states A and B yields different output; thus, A and B are distinguishable. Similarly, A and D are distinguished for input 0. C and B are distinguishable, as are C and D. (Why?) For any input sequence of length 1, A and C have the same output. However, the input sequence 00 yields different results for A and C: when we begin in state A with input 00, we end with output sequence 00, but beginning in C with 00 yields 01. Thus, all states

State	Next state Input		Output	
	0	1	0	1
A	B	C	0	0
B	D	B	0	1
C	D	C	1	0
D	D	A	1	1

Table 5.1

are distinguishable from one another, so we cannot reduce this table to a smaller one.

For a second example, look at Table 5.2. It is clear by examining the output columns that A is distinguishable from B and D. However, A, C, and E are indistinguishable by any sequence of length 1. Thus, we check sequences of length 2. Table 5.3 exhibits the output for all possible input combinations of length 2.

Note that A and C are distinguished by an input sequence of 11, A and E by input of 01 or 11, and C and E by an input of 01. However, B and D cannot be distinguished by a sequence of length 2. Must we check all possible sequences of length 3? Fortunately, the answer is no. First, we see that an input of 0 takes B to state E but takes D to state A. Since E and A are distinguished (as we showed above), there is a sequence x that distin-

State	Next state Input		Output	
	0	1	0	1
A	B	C	0	0
B	E	B	0	1
C	D	E	0	0
D	A	D	0	1
E	C	E	0	0

Table 5.2

Input sequence	Output Starting state				
	A	B	C	D	E
00	00	00	00	00	00
01	01	00	01	00	00
10	00	10	00	10	00
11	01	11	00	11	00

Table 5.3

guishes A from E. We can then use the sequence $0x$ to distinguish B from D. In this case, the sequence 01 distinguishes B from D (producing output sequences 001 and 000). In general, if any pair of states maps onto a second pair of distinguishable states for any input, then the first pair of states is also distinguishable.

Our final example (shown in Table 5.4) is considerably more complex. In this example, states A and B are distinguishable, since for input 1, the output is different. Similarly, states A and D are distinguishable. Even though A and D have the same output whatever the input, we get different output sequences for the same input sequence: for the input sequence 01, we get the output sequence 01 when we start from A and the output sequence 00 when we start from D. However, states E and H are indistinguishable.

Given a state transition diagram, we will look at each pair of states to see whether it is distinguishable or not. Such a technique allows us to obtain an equivalent reduced machine. That is, if two states are indistinguishable, they perform the same work; thus, they can be represented by just one state.

We partition the states in the machine into equivalence classes: two states are in one equivalence class if and only if the two states are indistinguishable. As we noted in the example, two states with different outputs are always distinguishable. Fortunately, there is an easy way to tell when two states are distinguishable. Suppose B and E are indistinguishable. Then, they must be in the same equivalence class. For an input of $x = 0$, both B and E must go to the same equivalence class (since two states are in different equivalence classes only if they are distinguishable). Suppose $B - 0 \rightarrow X$ (that is, from state B, an input of 0 sends us to state X) and $E - 0 \rightarrow Y$, where X and Y are distinguishable. Then, there must be a sequence of

State	Next state Input		Output	
	0	1	0	1
A	B	E	0	0
B	B	A	0	1
C	G	C	1	0
D	D	A	0	0
E	C	F	0	1
F	A	A	1	0
G	D	H	1	1
H	C	F	0	1

Table 5.4

inputs x_1, x_2, ..., x_n that distinguishes X from Y. Thus, the sequence 0, x_1, x_2, ..., x_n would distinguish B and E. Consequently, for B and E to be in the same equivalence class, they must both map onto identical equivalence classes for all inputs. If for some input, B and E map to two different equivalence classes, then B and E are distinguishable.

Let's see how this helps us to decide whether B and E are distinguishable. Suppose they are not. Then, B and E would be in the same equivalence class, denoted by BE. This equivalence class is also a block of a partition, since the equivalence relation divides all of the states into nonoverlapping blocks, and each state must be in exactly one block. However, if B and E are in the same block, so, too, must B and C be in one block and A and F in one block; that is, BC and AF are blocks of the partition. This is because B $- 0 \rightarrow B$ and $E - 0 \rightarrow C$, and $B - 1 \rightarrow A$ and $E - 1 \rightarrow F$. However, the blocks are nonoverlapping, so this means that B, C, and E must all be in the same block. So far, then, our blocks are BCE and AF. But a closer examination of B and C shows us that B and C are distinguishable, so they cannot be in the same block. We have arrived at a contradiction by assuming that B and E were indistinguishable, so it must be the case that B and E are distinguishable.

Next, let us look at C and F. If CF were a block, we would have blocks CF, AG, and CA. In other words, there would have to be one large block ACFG. We know, though, that A and C are distinguishable, so it cannot be the case that they reside in the same block. Therefore, C and F must be distinguishable. Continuing in the same way, we find that the blocks of this machine are

A   B   C   D   EH   F   G

We have thus reduced the size of the machine under consideration, and we can consider these blocks to represent the reduced machine. We can then rewrite the machine's table as Table 5.5.

State	Next state Input		Output	
	0	1	0	1
A	B	EH	0	0
B	B	A	0	1
C	G	C	1	0
D	D	A	0	0
EH	C	F	0	1
F	A	A	1	0
G	D	EH	1	1

Table 5.5

In essence, what we do to reduce a machine is try combinations of states with identical outputs to see whether they might be in the same block of the partition. The algorithm to decide which states are indistinguishable (i.e., one which generates the blocks of the partition) can be described in the following steps:

1. Let S_{ij} be the set of all states with output (i, j), where i and j are equal to 0 or 1.
2. For each S_{ij}, perform the following: For each pair of states X and Y in S_{ij}, suppose X and Y are not distinguishable. Then, XY is an equivalence class. Use this information and the state transition table to generate the remaining blocks of the partition. If a contradiction arises, then X and Y are distinguishable. If there is no contradiction, then X and Y are indistinguishable.

By putting the combinations into blocks and examining the consequences of doing so, we eliminate all possibilities except the final correct configuration of blocks. This method is sometimes time−consuming, but the results can make a substantial difference in the size of the resulting machine.

Let us work through another example to see more clearly how this method works. Suppose we are given Table 5.6 as a state transition table. Suppose AB is a block. This forces F and H into a block, and B and B (which is just B) into another. FH, in turn, forces DE and HF (which is the same as FH). DE forces blocks AA and FH to be formed. To summarize, this means that assuming AB is a block forces FH and DE to be blocks, too. Hence, AB, C, DE, FH, G is a partition that represents an equivalent reduced machine.

We will continue to examine other possibilities, since this may not be the smallest reduced machine we can find. Suppose that, instead of A and B being in a block together, we have A and C in the same block. Then, we are

State	Next state Input		Output	
	0	1	0	1
A	F	B	0	0
B	H	B	0	0
C	A	D	0	0
D	A	F	0	0
E	A	H	0	0
F	D	H	1	1
G	B	C	1	1
H	E	F	1	1

Table 5.6

forced to have BD and AF as blocks. However, A and F have different outputs, so they cannot be together in a block. This means that we cannot group A and C together. In a similar fashion, we try to pair A with the other states of the machine. Pairing A with D to form a block forces us to pair A with F, which we have shown is not possible. Likewise, pairing A with E also forces us to put A with F, so this pairing is unacceptable. Continuing with our analysis, we see that pairing A with either of F, G, or H in a block is impossible. Now we turn to B. Trying to form blocks BC, BD, or BE all force us to create block AH, and similar problems arise when BF, BG, and BH are tried. Putting C and D in the same block forces us to create block DF, and putting C with E yields block DH; both results are not possible. Combining C with F, G, or H is unsuccessful, too. When we reach the case where D and E are put in the same block, we find that DE forces the pairing of A with A and of F with H. FH mandates the pairing of D with E and F with H, so we have a reduced machine represented by

$$A \quad B \quad C \quad DE \quad FH \quad G$$

This partition has more states that the other reduced machine we found (by pairing A with B).

In order to exhaust all of the cases, we pair FG, only to find that this forces HC to be a block, which is not possible. When we form block FH, we end up with the same partition as the one generated by DE. Finally, forming block GH results in the formation of CF, which is also not possible. Thus, we have considered every case. (In the next chapter, we learn techniques for counting the number of cases to consider, so that you'll know what you are in for when you start applying this method.) The analysis has shown us that state A can only be in a block with state B, and vice versa; C cannot be in a block with anything else; D and E can be together in a block, as can F and H; and G must be in a block by itself. Thus, our reduced machine is as shown in Table 5.7.

As you can see, this machine is substantially smaller than the original

State	Next state Input		Output	
	0	1	0	1
AB	FH	AB	0	0
C	AB	DE	0	0
DE	AB	FH	0	0
FH	DE	FH	1	1
G	AB	C	1	1

Table 5.7

machine. Exercise 6 will give you a chance to put together some of the methods we have described here; you can transform a state diagram and state assignment table into a reduced machine.

MACHINE DECOMPOSITION

In the previous section, we showed how to analyze a state transition table in order to reduce the machine to a smaller one. Now we consider another question posed in Chapter 4: can we split a machine into two or more simpler machines? Recall that a machine might be decomposed into a serial or a parallel decomposition. Thus, it may start out looking like

and can be broken up as

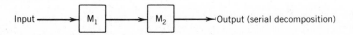

or

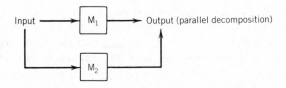

Why is a decomposition useful? It is easy to see that if a machine can be decomposed into two or more machines which can perform in parallel, then we can speed up the work done by the original machine. A serial decomposition may not result in an increase in speed, but it usually reduces the complexity of the original machine. For example, a six—state machine requires three flip—flops for implementation. However, if it is decomposed serially into a two—state machine followed by another two—state machine, then we can implement the serial decomposition with only two flip—flops (one flip—flop for each machine in the decomposition).

To achieve such decompositions, we define a lattice of partitions. The method we will employ to decompose machines is similar to that used when we reduced machines. However, in this case, we ignore the outputs. First, let us look at what are known as substitution property (or SP) partitions of the states of a machine. An **SP partition** is a partition of the states of a machine such that all states in a partition block, when presented with a given input symbol, proceed to states which are also together in a partition block. For example, the previous state reduction example yielded a partition of

AB   C   DE   FH   G

This is an SP partition. (Verify this for yourself.) Unlike the restrictions in the state reduction method, we could put A and F together in a block. (Why?) Let us look at how we can generate all possible SP partitions. By generating SP partitions and ordering them so as to form a lattice, we can use the properties of this lattice to decompose the machines.

ORDERING PARTITIONS

For any state table, we can consider the entire set of partitions of that state table. We know that the set of partitions always has at least two elements. One is a partition in which each state is in a block by itself. Thus, for a machine with states A, B, C, D, E, F, G, and H, the partition would look like

A   B   C   D   E·   F   G   H

This corresponds to the 0 partition of the lattice which contains all of the SP partitions. The 1 partition of the lattice is the one in which all of the states are in one block. Then, the partition would look like

ABCDEFGH

In such a lattice, we have an ordering of the partitions. We say that (partition X) $<$ (partition Y) if every block of X is included in a block of Y.

How can we generate other partitions in the lattice? Knowing one or more partitions, we can build other partitions from them. For example, we can have one partition operate on another. If Π_1 and Π_2 are two partitions with blocks

$$\Pi_1 = B_{11}, B_{12}, ..., B_{1n}$$

and

$$\Pi_2 = B_{21}, B_{22}, ..., B_{2m}$$

then we can define the operation ● on Π_1 and Π_2 as producing the partition resulting by intersecting each block of Π_1 with each block of Π_2. Thus, the blocks of the new partition $\Pi_1 \bullet \Pi_2$ would include

$$B_{1i} \cap B_{2j}$$

for all possible combinations of i and j. This is denoted as

$$\{ B_{1i} \cap B_{2j} \mid 1 \leq i \leq n, 1 \leq j \leq m \}$$

Let us look at an example. Suppose we have a set $\{ A, B, C, D, E, F \}$, and there are two partitions of this set:

$$\Pi_1 = \{ ABCD, EF \}$$

and

$$\Pi_2 = \{ ABF, CD, E \}$$

We must look at the intersection of each of the two blocks of Π_1 with each of the three blocks of Π_2, so there are six intersections to compute:

1. ABCD $\cap$ ABF $=$ AB
2. ABCD $\cap$ CD $=$ CD
3. ABCD $\cap$ E $=$ ϕ
4. EF $\cap$ ABF $=$ F
5. EF $\cap$ CD $=$ ϕ
6. EF $\cap$ E $=$ E

This means that the partition $\Pi_1 \bullet \Pi_2$ is $\{ AB, CD, E, F \}$. Let us look at another example. Let Π_3 and Π_4 be partitions of $\{ A, B, C, D, E \}$ defined by

$$\Pi_3 = \{ ABC, DE \}$$

and

$$\Pi_4 = \{ A, BE, CD \}$$

Compute $\Pi_3 \bullet \Pi_4$ by examining each of the possible intersections:

1. ABC $\cap$ A $=$ A
2. ABC $\cap$ BE $=$ B
3. ABC $\cap$ CD $=$ C
4. DE $\cap$ A $=$ ϕ
5. DE $\cap$ BE $=$ E
6. DE $\cap$ CD $=$ D

In this case, we see that $\Pi_3 \bullet \Pi_4$ is the partition { A, B, C, D, E }. The partitions Π_3 and Π_4 have a special property: *orthogonality*. You will remember that the partition { A, B, C, D, E } is the 0 partition in the lattice of partitions. We say that two partitions are **orthogonal** if the intersection of any block of one with any block from the other contains at most one element. In other words, two partitions Π_1 and Π_2 are orthogonal if and only if $\Pi_1 \bullet \Pi_2 = 0$.

Note that two partitions are orthogonal if each block of the second contains at most one element from any block of the first. This gives us a way of constructing a new partition from a given partition. We create this second, orthogonal partition by taking each block of the first and parcelling its elements out to different blocks of the second. For example, if Π is the partition

AC   BD   E

then we must put A in one block of the new orthogonal partition and C in another. Likewise, B and D must go into different blocks. By putting A and B together to form block AB, and likewise forming block CD, we can form the orthogonal partition AB, CD, E.

In general, if Π_1 is orthogonal to Π_2, and the largest block of Π_1 contains k elements, then any orthogonal partition must have at least k blocks.

GENERATING SP PARTITIONS

If X and Y are SP partitions, then so are $X \bullet Y$ and $X + Y$. (You will prove this in Exercise 13.) How do we generate all of the SP partitions of a machine? Doing so will allow us to generate all of the nodes of the lattice of SP partitions. Let us work through an example. Consider the state transition table shown as Table 5.8.

| | Next state | | | Output | |
| | Input | | | | |
State	0	1		0	1
A	D	A		0	0
B	E	A		0	1
C	E	C		1	1
D	B	G		0	0
E	C	H		1	0
F	D	A		1	0
G	A	G		0	0
H	C	H		1	1

Table 5.8

We pair states until we generate all possible implied SP partitions.

1. If AB is a block, then DE must be one, since $A - 0 \rightarrow D$ and $B - 0 \rightarrow E$. This implies, in turn, that BC and GH must be blocks. If BC is a block, then AC is one, and block GH yields blocks AC and GH. Therefore, we have SP partition 1:

$$ABC \qquad DE \qquad GH \qquad F$$

2. If AC is a block, we must also have DE as a block. Pursuing this results in a partition identical to SP partition 1.
3. If A and D are in the same block, then BD and AG must be blocks. However, since blocks cannot overlap, this says that ABDG must be a block. From ABDG, we know that DEBA and AG must be blocks, but that just expands our existing block to ABDEG. This new block generates DEBCA and AGH, but that forces ABCDEGH to be a block. The large block ABCDEGH generates itself, so we have SP partition 2 defined by

$$ABCDEGH \qquad F$$

4. Trying to group A and E in a block results in forcing DC and AH to be blocks. This, in turn, forces blocks EB and CG from the existence of DC, and it also forces CD and AH from the existence of AH. Following this case to its end, we generate a second copy of SP partition 2.
5. If A and F are in the same block of the partition, then D and A must each be blocks of the partition. This generates SP partition 3:

$$AF \qquad B \qquad C \qquad D \qquad E \qquad G \qquad H$$

6. Placing A and G in the same block generates block AD and thus ADG. From case 3, we know that AD leads to SP partition 2.
7. Were A and H to be in the same block, then CD would have to be a block. CD forces blocks EB and CG to be in the partition, so AH, CDG, and BE would be blocks. CDG forces block ABE, so we must have SP partition 2.
8. Block BC implies block AC, so ABC would be a block of this partition. However, AB leads to SP partition 1.
9. BD forces AG to be a block, and case 6 tells us that this is SP partition 2.
10. BE results in having a block containing A and H. Case 7 includes this block, and this leads to SP partition 2.
11. If B and F are in the same block of a partition, then ED must be another block. ED forces BC and GH to be blocks, so this far we have BCF, ED and GH. We know from case 8 that BC leads to SP

partition 1, which has DE and GH in blocks. However, block GH forces us to have AC as a block, and we end up with SP partition 4:

ABCDEFGH

12. Block CD implies that BE and CG must be blocks. However, having block BE and referring to case 10 shows us that this is just SP partition 2.
13. If C and E are together in a block, then CH must also be a block. We thus define SP partition 5:

CEH   A   B   D   F   G

14. Placing C and F in the same partition block means that AC and DE must be blocks. AC gives us SP partition 1, but we also have F in the ABC block, so SP partition 6 becomes

ABCF   DE   GH

15. Pairing C and H in a block forces C and E to be in a block; we end up with SP partition 5.
16. If D and E are together in a block, then BC and GH must be blocks. This is SP partition 1.
17. The block DF forces BD and AG to be blocks. AG leads to SP partition 2 with F in the same block as A. Thus, this case leads to SP partition 4.
18. DG results in AB as a block. AB, in turn, forces ABC, DE, GH, and F to be blocks. Since we must have D and G together, we get ABC, DEGH, and F as blocks. Block DEGH forces BCA and GH to be blocks, so the result is SP partition 7:

ABC   DEGH   F

19. If D and H are in the same block, then BC and GH are forced to be blocks. This means that DGH must be a block, and we have SP partition 7.
20. Block EF implies that CD and AH are blocks. However, AH leads to SP partition 2 with F added; this is SP partition 4.
21. EG results in the formation of blocks AC and HG, and the result is SP partition 7.
22. By placing E and H in the same block, we must have C and H as distinct blocks. This generates SP partition 8:

EH   A   B   C   D   F   G

23. If F and G are in the same block, then AD and AG must be blocks. This leads to SP partition 4.

24. When F and H are placed in the same block, CD and AH are forced to be blocks. This is also SP partition 4.
25. If GH is a block, then AC must be a block. The result is SP partition 1.

In this manner, we have generated the following SP partitions:

1. ABC DE F GH
2. ABCDEGH F
3. AF B C D E G H
4. ABCDEFGH
5. CEH A B D F G
6. ABCF DE GH
7. ABC DEGH F
8. EH A B C D F G

Recall that there is a ninth SP partition:

  A B C D E F G H

and we can form $X + Y$ and $X \bullet Y$ for all of these. We generate three more partitions in this manner:

10. AF CEH B D G (3 + 5)
11. ABCF DEGH (3 + 7)
12. AF EH B C D G (3 + 8)

We calculated all of these partitions because we want to look at the lattice of the partitions. It looks like Figure 5.13. In this lattice, partitions 4 (where all states are in one block) and 9 (where each state is in its own block) are the trivial partitions. Are there any nontrivial partitions that are orthogonal? By examining Figure 5.13, we see that the following pairs of partitions are orthogonal because the greatest lower bound of each pair is Π_9.

2 and 3	10 and 1	3 and 1
7 and 3	5 and 1	5 and 3
6 and 5	12 and 1	8 and 3
6 and 8	8 and 1	

The number of blocks in an SP partition is important because we can form a machine based on the number of blocks. Note that if SP partition Π_i is an upper bound for SP partition Π_j, then Π_i will have fewer blocks than Π_j. SP partition 1 of our example has four blocks:

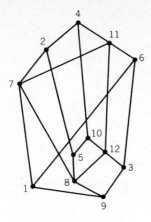

Figure 5.13

ABC   DE   F   GH

To form a machine from these blocks, we determine the next state as shown in Table 5.9. This new machine has four states, thus requiring only two flip—flops to represent it. As you can see, the fewer the number of blocks in an SP partition, the simpler the machine based on that partition.

FORMING A SERIAL DECOMPOSITION

We want to form a serial decomposition of a machine in order to reduce its complexity. Recall that two partitions are orthogonal if and only if their composition is the 0 partition. Let us use this property to find a serial decomposition of a machine. To form the serial decomposition, we must have an SP partition of the machine. We want to define two new machines from it whose action together is equivalent to the original machine. In the machine we studied in the last section, we developed the lattice of parti-

State	Next state	
	Input	
	0	1
ABC	DE	ABC
DE	ABC	GH
F	DE	ABC
GH	ABC	GH

Table 5.9

| | Next State | | |
State	0	1	Output assignment
ABCF	DE	ABCF	00
DE	ABCF	GH	01
GH	ABCF	GH	10

Table 5.10

tions. For example, we have the SP partition ABCF, DE, GH. We can use this partition to define the first machine as shown in Table 5.10. The output assignment in Table 5.10 serves only to identify the state of the first machine. The choice of the assignment is arbitrary as long as the assignment identifies uniquely the current state.

We can determine the second machine in the following manner. We find a partition Π (not necessarily an SP partition) having the property that it is orthogonal to the partition defining the first machine. Thus, we want to find a Π such that

$$\{ \text{ABCF, DE, GH} \} \bullet \Pi = 0 = \{ \text{A, B, C, D, E, F, G, H} \}.$$

There may be several partitions which satisfy that requirement. One possibility is the partition { ADG, BEH, C, F }. (You will verify in Exercise 14 that these partitions are indeed orthogonal.) This partition can be used to define the second machine. In Table 5.11, the input, x, is the input to the original machine. Inputs y and z are the output values of the 3$-$state machine shown in Table 5.10. Thus, when in the second machine, we can tell from input xyz the block from which we came in the first machine. The entries of dashes ($---$) indicate that these are "don't$-$care" states. To see how this machine was obtained, look at the ADG entry under column 000. If we are in the ADG block, we must be either in state A, state D, or state G. If the yz values are 00, we must have been in the ABCF block of the first machine. However, these machines were derived from orthogonal partitions, and ADG ∩ ABCF = A, so we must have been in state A. For an input of $x = 0$, we arrive at state D in the ADG block; for $x = 1$, we move to state A, also in the ADG block. Thus, the yz inputs from the first

xyz	0 00	1 00	0 01	1 01	0 10	1 10
ADG	ADG	ADG	BEH	ADG	ADG	ADG
BEH	BEH	ADG	C	BEH	C	BEH
C	BEH	C	---	---	---	---
F	ADG	ADG	---	---	---	---

Table 5.11

machine and the current state serve to identify uniquely the exact current state. In the same manner, we can generate the output.

In the resulting serial decomposition, the second machine contains information about the block in the first machine (as described above), and orthogonality enables us to determine uniquely the state in which the original machine must have been. Note that we had many nontrivial SP paritions from which to choose for building a serial decomposition. The original 8−state machine required 3 flip−flops for its representation. The two machines above into which we decomposed the original machine, al- though much simpler than the original and thus easier to construct, require 2 flip−flops each. Optimality can be achieved in the choice of machines for a serial decomposition, but it is beyond the scope of this text.

PARALLEL DECOMPOSITION

A parallel decomposition of a machine allows us to have at least two machines working simultaneously to perform the same work as the original singular machine. For parallel decomposition of a machine, we need two nontrivial (i.e., not 1 and not 0) SP partitions that are orthogonal. For example, the partitions { ABC, DE, F, GH } and { AF, CEH, B, D, G } are orthogonal. These two partitions define two machines M_1 and M_2 that (unlike serial decomposition) run completely independent of one another. If you know which block M_1 is in and which block M_2 is in, then you know which state the original machine is in, and you can determine the output accordingly. Table 5.12 shows an example of how machines M_1 and M_2 can be defined. As with the construction of machines for a serial decomposition, the output assignment is arbitrary. A final logic device (requiring no flip−flops) uses the output from M_1 and M_2 and the input to the original machine to determine the final output from the parallel machines.

From the machines M_1 and M_2, we can form a table of final output assignments. (See Table 5.13.) Each entry x/y consists of an original state

State	Input 0	Input 1	Output
ABC	DE	ABC	00
DE	ABC	GH	01
F	DE	ABC	10
GH	ABC	GH	11

M_1

State	Input 0	Input 1	Output
AF	D	AF	000
CEH	CEH	CEH	001
B	CEH	AF	010
D	B	G	011
G	AF	G	100

M_2

Table 5.12

M_1 output

		Input = 0				Input = 1			
		00	01	10	11	00	01	10	11
M_2	000	A/0	—	F/1	—	A/0	—	F/0	—
	001	C/1	E/1	—	H/1	C/1	E/0	—	H/1
output	010	B/0	—	—	—	B/1	—	—	-
	011	—	D/0	—	-	—	D/0	—	—
	100	—	—	—	G/0	—	—	—	G/0

Table 5.13

(x) and a final output value (y). The dashes indicate don't—care or illegal cases. To see how the table is formed, suppose the output from M_2 is 000. This output represents state AF in machine M_2, so we must have been either in state A or in state F in the original machine. If the output of M_1 is 00, then M_1 must have been in state ABC. Orthogonality assures us that the intersection of the two blocks (AF in M_2 and ABC in M_1) yields at most one state of the original machine; in this case, the result is state A, so we must have been in state A in the original machine. Finally, for input values of 0 and 1 in the original machine, we know the output from state A.

A similar analysis of other combinations allows us to fill in the remaining entries of Table 5.13. For example, if the output from M_2 is 000 and from M_1 is 01, we find that this is an impossible situation. In this fashion, two orthogonal nontrivial SP partitions result in a parallel decomposition of the original machine.

STATE ASSIGNMENT

The third of our three questions about state tables deals with state assignments. Let us look back at our example for generating SP partitions. The machine has eight states, requires three flip—flops, and thus must have three state assignment variables so that each state can be represented by a unique combination of variables. If we call the variables y_1, y_2, and y_3, we can assign 000 to state A, 001 to state B, and so on, ending with 111 as state H. The realization of this machine will require some number of gates. It is likely that another state assignment generates a different Karnaugh map with a different number of gates. There are 8! = 40,320 possible state assignments. (Why?) Let us see how we can find one that uses the smallest number of gates.

In an arbitrary assignment of variables to states, the next value of y_1 depends on the current values of y_1, y_2, y_3, and the input x. Using the SP partition $\Pi_{11} = \{$ ABCF, DEGH $\}$ as an example, suppose we choose y_1, thus $y_1 = 0$ for states A, B, C, and F, and 1 for states D, E, G, and H. If the input is 0, we move to the DEGH block; if the input is 1, we remain in the ABCF block. Thus, the use of an SP partition assures us that the new value of y_1 depends only on the old value of y_1 and the input x. By defining y_1 in terms of blocks of an SP partition, we get a simpler expression for y_1.

We can simplify the expressions for y_2 and y_3 in a similar way by finding two 2−block SP partitions Π_i and Π_j such that

$$\Pi_{11} \bullet \Pi_i \bullet \Pi_j = 0 = \Pi_9$$

Such an expression would assure us of unique assignments for each state. Unfortunately, no such Π_i and Π_j exist for this machine.

Next, we try to find an SP partition Π_i with at most four blocks and satisfying

$$\Pi_{11} \bullet \Pi_i = 0$$

By having at most four blocks, we can assign unique combinations of $y_2\, y_3$ to each block; thus, the new values of y_2 and y_3 are independent of y_1, and we can simplify the expressions for y_2 and y_3.

We have gained a great deal by using this approach since there are other partitions (i.e., not necessarily SP partitions) that will help. Let us see how. We say that Π_1 and Π_2 form a **partition pair** if the inputs map blocks of Π_1 onto blocks of Π_2. If $\Pi_1 = \Pi_2$, we have an SP partition. To be effective, though, Π_1 need not be equal to Π_2. Recall that in the last example, the new value of y_1 depended only on the previous value of y_1 and the input. With a partition pair, if we know the block of Π_1 we are in, then by examining the input we can determine the block of Π_2 we will be in. In general, given state assignment variables y_1, y_2, ..., y_k based on the blocks of Π_1 and variables y_r, y_{r+1}, ..., y_s based on the blocks of Π_2, the new values of y_r, ..., y_s will depend only on the input and the old values of y_1, ..., y_k. This allows us to develop simpler expressions for the new values of y_r, ..., y_s.

An example of this may be seen in Table 5.14. You can see that we have two partition pairs:

$$
\begin{array}{ll}
AB \quad CD \rightarrow AD\ BC & (\Pi_1 \rightarrow \Pi_2) \\
AD \quad BC \rightarrow AB\ CD & (\Pi_2 \rightarrow \Pi_1)
\end{array}
$$

	x	y	z
A	D	B	B
B	A	C	B
C	B	C	A
D	C	B	A

Table 5.14

y_1	y_2			
0	0	A	D	B	B
0	1	B	A	C	B
1	1	C	B	C	A
1	0	D	C	B	A

Table 5.15

Thus, blocks AB and CD map onto blocks AD and BC. Now, we make state assignments based on these pairs. For partition Π_1, we let y_1 be 0 in block AB and 1 in block CD. Likewise, for partition Π_2, we let y_2 be 0 for block AD and 1 in block BC. This yields Table 5.15.

The table shows us that y_1 is a function of y_2 and the inputs, while y_2 is a function of y_1 and the inputs. This says that we are not limited to SP partitions; we can add to the SP partitions in order to achieve optimal state assignments.

The example above merely begins to explore the intricacies of the state assignment problem. In the general case, we must be concerned with three classes of assignments:

1. Input assignments
2. State assignments
3. Output assignments

Consider an example where the assignment table looks like Table 5.16. In this example, we need three state variables, two input variables, and two output variables. The two output variables can generate four possible output values. To achieve an optimal realization, assignments must be made to all variables in an appropriate fashion. A full analysis of this complex a problem is beyond the scope of this text. For more information, see Hartmanis and Stearns, *Algebraic Structure Theory of Sequential Machines* (Prentice-Hall, 1966), or Kohavi, *Switching and Finite Automata Theory* (McGraw-Hill, 1978).

	X	Y	Z	Outputs		
A	B	E	D	j	k	j
B	C	C	F	j	l	k
C	A	D	C	l	l	m
D	B	A	A	m	j	k
E	F	E	C	m	m	m
F	A	B	C	k	l	j

Table 5.16

This chapter has introduced you to a number of concepts, all of which are related to state assignments and the realization of assignment tables as machines. The exercises will assist you in putting together the various pieces which were introduced in this chapter.

EXERCISES

1. The state diagram in Figure 5.14 and accompanying assignment table describe a machine. Using the techniques described in this chapter, rewrite the state assignment table, determine which states are indistinguishable from the others, and reduce the number of states in the machine.

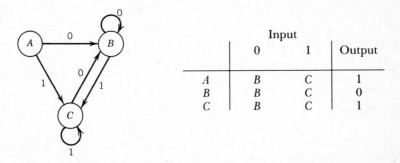

| | Input | | |
	0	1	Output
A	B	C	1
B	B	C	0
C	B	C	1

Figure 5.14

2. The state diagram in Figure 5.15 and accompanying assignment table describe a machine. Using the techniques described in this chapter, rewrite the state assignment table, determine which states are indistinguishable from the others, and reduce the number of states in the machine.

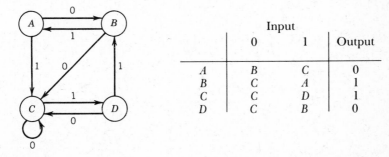

| | Input | | |
	0	1	Output
A	B	C	0
B	C	A	1
C	C	D	1
D	C	B	0

Figure 5.15

3. The state diagram in Figure 5.16 and accompanying assignment table describe a machine. Using the techniques described in this chapter, rewrite the state assignment table, determine which states are indistinguishable from the others, and reduce the number of states in the machine.

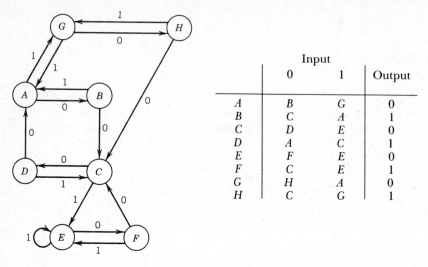

	Input		
	0	1	Output
A	B	G	0
B	C	A	1
C	D	E	0
D	A	C	1
E	F	E	0
F	C	E	1
G	H	A	0
H	C	G	1

Figure 5.16

4. For the table below, find the lattice of SP partitions. Find (if possible) a serial decomposition and a parallel decomposition.

	0	1
A	B	F
B	A	D
C	C	E
D	B	C
E	A	A
F	C	B

5. Find the partition pairs for the table below:

	x	y	z
A	A	B	D
B	C	D	B
C	B	A	C
D	D	C	A

6. Reduce the following machine.

	Input		Output	
	0	1		
A	D	F	0	0
B	C	A	0	0
C	A	E	0	1
D	A	B	0	0
E	B	F	1	0
F	C	A	0	0

7. Let A be the set { a, b, c, d, e } and R the relation { (a, a), (b, b), (c, c), (d, d), (e, e), (a, b), (b, d), (a, d) }. Verify that R is a partial order on A. Is this a lattice? Why or why not?

8. For the Hasse diagram in Figure 5.4, enumerate the pairs of the relation that defines the partially ordered set. What is the dual of this partially ordered set? Draw the Hasse diagram for the dual.

9. Let S be the set of positive integers greater than 1, and let R be the relation defined by xRy if and only if x is a divisor of y. Does the partially ordered set (S, R) have a supremum? What is the dual of (S, R)?

10. Prove Theorem 5.2.

11. Prove Theorem 5.4.

12. Show that lattices F and G defined in the section on distributive lattices are indeed distributive.

13. Show that if X and Y are SP partitions, so are $X \bullet Y$ and $X + Y$. This means that the set of SP partitions is closed under the operations $\bullet$ and $+$.

14. In this chapter, we defined a machine by the following partition:

	Input		Output assignment
	0	1	
ABCF	DE	ABCF	00
DE	ABCF	GH	01
GH	ABCF	GH	10

Show that this partition is orthogonal to { ADG, BEH, C, F }.

15. How many flip—flops are needed to represent three states? Eight states? Twelve states? Twenty—five states? Two hundred states? How did you compute each of these answers?

CHAPTER 6

CARDINALITY AND COUNTABILITY

In previous chapters, we dealt with issues of *size*. In a nested loop within a loop, for instance, we needed to have some idea of how many times our program would increment a variable. When generating a truth table, it was important to enumerate all of the cases; that is, the number of variables in the truth table determined the number of rows in the table. When we drew a Hasse diagram of a lattice, we wanted to be sure that we had considered all of the possible pairs of nodes when drawing the relationship among the points of the lattice. Thus, in many instances, it has been important for us to know something about the size of the set we are dealing with or the size of the problem we are about to tackle. It is for this reason that in this chapter we turn to the notion of the size of a set.

CARDINALITY OF A SET

When we think of the size of a set, we are really thinking about the number of elements in that set. For instance, the set S of letters of the English alphabet has 26 elements, so we say that the size of S is 26. Note that size is a property of the set S. We may denote the size of S by writing 'twenty–six' or '26' or 'xxvi' or '11010' but we are still describing the same property of S. We call the number of elements of a set A the **cardinality** or **size** of A, and we denote it by $| A |$. Thus, we can write $| S | = 26$.

Is the cardinality of a set something which never changes? Consider the set of residents of the state of Illinois. As people move in and out of the state, as people are born and others die, the number of residents changes. Does this mean that the cardinality of the set changes? The cardinality of this set is dependent on the point in time at which it is measured; the set's cardinality last month is likely to be different from its cardinality next month. The cardinality changes only when the set's members change. For a set at a given time, the cardinality is unchanging. Problems in measuring the cardinality of a set arise in finding effective methods for counting the set's elements. For example, today it is impossible to measure the cardinality of the set of students who will be taking Discrete Structures at the University of Colorado in 1999. Thus, sets have cardinalities, independent of whether we have methods for counting the number of their elements.

How do we determine the cardinality of a set? As described above, we can count the elements of the set, if possible. When counting is not a reasonable option (for instance, if the set is not finite), we can compare the given set to a set whose cardinality is known to see whether the cardinalities of the sets are the same. For example, we can ask whether there are more integers than even integers. Are there more odd integers than even integers? Are there more real numbers than integers?

In much of our daily activity, we compare the cardinalities of two sets to see whether they are equal or one is less than the other. In setting the dinner table, we want the cardinality of the set of plates on the table to be equal to the cardinality of the set of people who will be at dinner. Likewise, in ordering textbooks for next semester's Discrete Structures course, the bookstore manager wants to be sure that the cardinality of the set of textbooks on the shelves is at least as large as the cardinality of the set of students enrolled in Discrete Structures. What we are doing when we compare the sizes of sets is matching up those sets, element for element. In other words, for each plate, there must be exactly one diner, and for each student, there must be one book. Mathematically, this is the equivalent of finding a one−to−one function which maps elements of the first finite set to elements of the second finite set. If we use up all the elements of the first set before we use up the elements of the second set, then we say that the first set is smaller than the second. If, on the other hand, we use up all elements of the second set and still have elements of the first, then we say that the first set is larger in size or cardinality than the second. Finally, if we complete our matching and find that we have no elements of either set left over, then we say that the sets are of the same size.

More formally, let A and B be finite sets. If there is a one−to−one function which maps A onto B, then we say that

$$| A | = | B |.$$

If, however, there is a one−to−one function from A onto a proper subset of B, then we say that

$$| A | < | B |.$$

INFINITE SETS

Does this method of comparison work only with finite sets? No. We can also compare infinite sets to each other or to finite sets by establishing the appropriate one—to—one mappings. In fact, we use the set of positive integers as a standard for comparison for some sets. We say that a set S is **countable** or **denumerable** if there is a one—to—one mapping from S onto the set Z of positive integers. We say that we can number the elements of a countable set because of the one—to—one function that relates each set element to a unique number. Please note that for infinite sets, the existence of a one—to—one function that is *not* onto does *not* preclude the existence of one that *is*. For instance, we can map the nonnegative integers to themselves by the mapping

$$f(n) = n + 1$$

This mapping is one—to—one but not onto, since there is no nonnegative integer which is mapped to 0 by f. However, the mapping

$$G(n) = n$$

is one—to—one and onto. Let us look at some other examples. Is the set E of all even positive integers a denumerable set? Let us define a mapping from E to Z in the following way:

For each x in E, define $f(x) == x//2$.

Check for yourself that f indeed is a one—to—one mapping from the set E = { 2, 4, 6, 8, ... } to the set Z = { 1, 2, 3, 4, ... }. This should not surprise you, for it simply implies that we can list or enumerate the elements of E in an orderly manner.

We can picture the function f as in Figure 6.1. Remember that with a one—to—one mapping from one set onto another, we also have an inverse

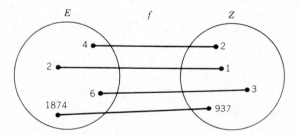

Figure 6.1

mapping. We can show that any element of E is associated with a unique element of Z. Also, every element of Z is the image of some element of E. This is clear when examining Z and E; the general concept of assigning a number to each element of a set is very useful if we are trying to process all elements of a set as, for example, we might do to take a census of the residents of Illinois. Note that the set E is a proper subset of the set Z, and yet the sets have the same cardinality. This seeming contradiction is exactly what distinguishes infinite sets from finite ones. An **infinite** set is equivalent in cardinality to some proper subset of itself. All other sets are **finite**.

Suppose that a hotel has a infinite number of rooms. Knowing that there is a bijection from the set of rooms to a proper subset of the set of rooms, we can conclude that the hotel can accommodate infinitely many new guests, even when the hotel is full.

Because we frequently compare other infinite sets to the cardinality of the set Z, the mathematician Cantor assigned a special symbol to the cardinality of Z. We write the cardinality of Z as $\aleph_0$. This is pronounced "aleph−zero," "aleph−null," or "aleph−nought," where the aleph is the first character in the Hebrew alphabet. We will soon see that not all infinite sets have the same cardinality, so that Cantor's $\aleph_0$ is the first in a series of **transfinite cardinal numbers** (continuing as $\aleph_1, \aleph_2, \aleph_3, \aleph_4$, and so on) used to describe the sizes of infinite sets.

Some other surprising results await us. Let T be the set of nonnegative integers. Then T is the same as the set Z united with the singleton set { 0 }:

$$T = Z \cup \{ 0 \}$$

We can define a one−to−one mapping from T onto Z by letting

$$f(x) = x + 1 \text{ for each } x \text{ in } T$$

Figure 6.2 shows how T and Z are related. Thus, | T | is also $\aleph_0$. In fact, we can show that the set I of *all* of the integers is of cardinality $\aleph_0$. The function we use assigns 0 to the 0 in the set T defined above, associates the negative numbers with the even positive integers, and associates the positive numbers with the odd positive integers. Formally, this becomes the function $g: I \rightarrow T$ where

$$g(x) = 0 \text{ if } x = 0$$
$$g(x) = -2x \text{ if } x < 0$$
$$g(x) = 2x - 1 \text{ if } x > 0$$

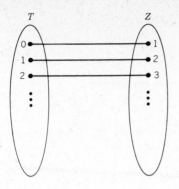

Figure 6.2

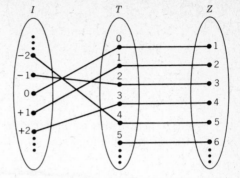

Figure 6.3

Figure 6.3 depicts how g relates I and T. Verify for yourself that g is a one−to−one mapping of I onto T. This means that $\mid I \mid = \mid T \mid = \aleph_0$.

We can also show that the set of fractions a/b, where a and b are both nonnegative integers, is also of cardinality $\aleph_0$. However, it is a bit more complicated to describe the one−to−one correspondence. We do this by proving a general result about denumerable sets. Then, we can apply this result to the set of fractions. The technique is a handy one that is applicable to many different situations. First, we prove the following theorem, which was first proved by Cantor.

THEOREM **6.1** _____

If the sets A_i are denumerable, for $i = 1, 2, ..., n$, then

$$\overset{n}{\underset{i=1}{\cup}} A$$

is denumerable.

_____

To prove this, let us number the elements of each of the sets A_i. (We can do this because each A_i is denumerable.) We label the elements of A_i as $a(i, 1), a(i, 2), a(i, 3), ..., a(i, k), ...$ Thus, $a(i, k)$ is the kth element of the set A_i. We can define the **height** of an element $a(i, k)$ to be the sum $i + k$. We then have no elements of height 1, only one element of height 2 (namely, $a(1, 1)$), two elements of height 3 ($a(1, 2)$ and $a(2, 1)$), and so on. For any positive integer n, there will be $n-1$ elements of height n, so we can arrange the elements of the union $\cup\ A_i$ in order by height. Thus, we have

$$\overset{\longleftarrow 2 \rightarrow}{}\ \overset{\longleftarrow \quad 3 \longrightarrow}{}\ \overset{\longleftarrow \qquad\qquad 4 \qquad\qquad \longrightarrow}{}\ \overset{\longleftarrow\qquad\qquad 5 \qquad}{}-\dots$$
$$a(1,\ 1),\ a(2,\ 1),\ a(1,\ 2),\ a(3,\ 1),\ a(2,\ 2),\ a(1,\ 3),\ a(4,\ 1),\ \dots$$

We can picture this as aligning the sets A_i one below the other, and we count the elements by following the arrows:

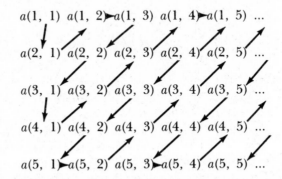

Clearly, this algorithm counts all the elements in the union, so the union must be denumerable. In other words, the arrows describe how we would define our one—to—one function: $a(1,\ 1)$ maps to $1, a\ a(2,\ 1)$ maps to 2, $a(1,\ 2)$ maps to 3, $a(1,\ 3)$ maps to 4, $a(2,\ 2)$ maps to 5, and so on. ■

As a consequence of this, we can derive the following result.

COROLLARY 6.2

The set of all rational numbers is denumerable.

Recall that the set of rational numbers is the set of numbers of the form a/b, where a and b are integers. We can define the set S_n to be the set of such fractions with denominator n. In other words, $S_n = \{\ j\ /\ n\ |\ j$ is an integer $\}$. It is clear that each set S_n is of cardinality $\aleph_0$ (that is, it is denumerable), so Theorem 6.1 assures us that the union of the S_n (i.e., the set of rational numbers) is countable. ■

TRANSFINITE CARDINAL NUMBERS

With infinite sets, we have seen that one set can be a proper subset of another while still having the same cardinality. Let us see when two infinite sets have different cardinalities. When we examined the power set of a finite set, we proved that if a finite set has n elements, its power set has 2^n elements. Thus, if A is a finite set, the cardinality of A (namely, n) is smaller than the cardinality of the power set of A (that is, 2^n). The same thing is true for infinite sets. If A is an infinite set, then $|A| < |P(A)|$. We can see why by proving the following theorem.

THEOREM 6.3 _____

For any nonempty set A, there is no one—to—one mapping from A onto the power set of A.

We will prove this theorem by contradiction. Thus, we will assume that there is such a one—to—one correspondence, and we will show how this assumption leads to a contradiction.

Let $f:A \rightarrow P(A)$ be a one—to—one mapping from A onto P (A). Then, for every element B in $P(A)$, there must be an element b in A such that $f(b) = B$, since f is an onto mapping. Because B is a subset of A, the element b is either an element of B or it isn't.

Define a set C in the following way:
$C = \{\ a \in A \mid a$ is not an element of $f(a)\ \}$

This set C may be empty, but it is a subset of A nevertheless. Thus, C is an element of the set $P(A)$. Because f is an onto mapping, there must be an element of A which maps to C; call this element x. If x is an element of C, then by the definition of C, x cannot belong to C. Likewise, if x is not an element of C, then the definition of C mandates that x be placed in C. Thus, either case is impossible: x can neither be in C nor out of it. This says that C is in the power set of A, but there is no element of A which is mapped to C under f. Therefore, our original assumption that such a function f exists must be false. ■

The proof of this theorem resembles the discussion in Chapter 2 of Russell's Paradox. After defining a set, we show that it is impossible for an element to be in the set and equally impossible for an element not to be in the set. Such a proof technique is common.

Note that this theorem makes no assumptions about whether A is finite. This result is not too surprising for finite sets, since the power set of any nonempty set is larger than the set itself. For infinite sets, however, our intuition about size may be misleading. This theorem also tells us that we can have infinite sets of differing cardinalities. We defined $\aleph_0$ as the cardinality of Z, the set of positive integers. This means that the cardinality of P(Z) is larger than the cardinality of Z. In other words, the cardinality of the power set of Z is greater than $\aleph_0$. We define $\aleph_1$ as the cardinality of the power set of Z. Sometimes, $\aleph_1$ is written as $2^{\aleph_0}$. We can continue in the same fashion to form the power set of the power set of Z and get an even larger infinite number: $\aleph_2 = 2^{\aleph_1}$. It is easy to see that this is an endless process, and we can use it to generate all the transfinite numbers.

It is reasonable to wonder, as many mathematicians did, whether there are any numbers between the transfinite numbers. In other words, is it possible to have a number x such that

$$\aleph_i < x < \aleph_{i+1}\text{s?}$$

The **continuum hypothesis** states that the answer to this question is no. In the early 1960s, Paul Cohen, then a student at Stanford University, proved that the continuum hypothesis is independent: it can be proved neither true nor false using the rules of set theory.

REAL NUMBERS

You may have noticed that we have talked only of integers; there has been no mention yet of the real numbers. Let us turn to the real numbers, then, and see how they relate to the cardinalities we have already introduced. Without knowing it, you often use a technique that shows exactly how the real numbers can be written in terms of the integers. In grade school, you learned how to work with decimal points and decimal fractions. You could write any number as either a terminating or nonterminating decimal number. Thus, 1/3 is the same as .3 (where the 3 repeats forever); 3/8 is also written as .375000 ... (where the 0 repeats forever); and π is the nonrepeating 3.141592653589 .... We will use this property to prove the following theorem.

THEOREM 6.4 _____

There is a one−to−one mapping from the set of subsets of the positive integers onto the set $Q = \{\ x$ a real number $|\ 0 \leq x < 1\ \}$.

___

Just as we can write any real number as a decimal expansion, we can also write that number as a binary expansion. This means that for any real number between 0 and 1, we can write that number as

$$. \, d_1 \, d_2 \, d_3 \, d_4 \, ... \, d_k \, d_{k+1} \, ...$$

where each of the d_k is a single binary digit, either 0 or 1. If this number has a terminating representation, then for some j, $d_k = 0$ for all $k > j$. Let us define the one−to−one correspondence with the subsets of the integers in the following way. With each real number x in the set Q, we will associate a set S_x of integers. The integer k belongs to the set S_x if and only if digit d_k is 1 in the binary expansion of x. Thus, each binary number uniquely defines a subset of the integers. Conversely, every subset of the integers corresponds to a unique binary expansion. For example, the expansion .011010000 ... maps to the set $\{\ 2,\ 3,\ 5\ \}$, while $0 =$ .00000000 ... maps to the empty set. Every real number in Q has a unique binary representation. Note that the real number $\pi - 3.0 = 0.14159\,...$ has no repeating representation in either decimal or binary notation. (See Exercise 1.) Nevertheless, the binary notation for $\pi - 3.0$ gives rise to a single set from P(Z). ■

___

This theorem tells us that the cardinality of the set Q is $\aleph_1 = 2^{\aleph_0}$. Thus, the cardinality of Q is greater than the cardinality of the set of integers. In terns of countability, this says that we can't enumerate the real numbers between 0 and 1.

Does it matter that we chose the interval between 0 and 1? Not really, because we can map any interval of real numbers to the interval between 0 and 1 (or to any other interval of real numbers) by drawing the intervals in parallel as shown in Figure 6.4. By picking a point off both of these intervals (such as the point p indicated), we can define a one−to−one correspondence f by drawing a line from p through both intervals. Each line radiating from p uniquely pairs a point on one interval with a point on the other interval. Thus, if a line through p crosses the smaller interval at point x, then $f(x)$ is the point where this same line crosses the other interval. This

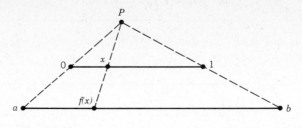

Figure 6.4

means that any bounded interval of real numbers has the same cardinality as any other interval of real numbers. Unfortunately, this implies that not only can we not count the real numbers between 0 and 1, but we cannot count any interval of real numbers.

COMBINING FINITE SETS

Let us return to finite numbers and finite sets to see what happens when we form the union of finite sets. If A is a finite set, then there is an integer n for which $|A| = n$. In this case, we refer to A as an **n−set**. If P is a proper subset of A with k elements, we call P a **k−subset** of A. Suppose we have two sets, A and B, where A is an n−set and B is an m−set. We want to know what happens if we put A and B together; how many elements do we have? It is very easy to see that if A and B have no elements in common, then $A \cup B$ has $m + n$ elements. More formally, we call this the

RULE OF SUM: If A is an m−set and B is an n−set, and if $A \cap B = \phi$, then $A \cup B$ is an $(m + n)$−set.

(You will be asked to prove this in the exercises.) Suppose, instead of only two sets, we have a collection of sets. We can generalize the rule of sum to say that

GENERALIZED RULE OF SUM: If, for each $i = 1, 2, ..., t$, we have a set A_i which is an n_i−set, and if $A_i \cap A_j = \phi$ for all pairs of distinct i and j, then $\cup A_i$ is an $(n_1 + n_2 + ... + n_t)$−set.

The generalized rule of sum is just another way of saying that if S is a set which can be partitioned into a collection $\{A_1, A_2, A_3, ..., A_t\}$ of disjoint sets, then the number of elements in S is the sum of the number of elements of the sets of the partition.

What happens if we break a set into subsets, but those subsets overlap? Then, counting the elements in the set becomes a very difficult problem, because we must make sure that we don't count an element twice and that we don't leave out an element. We will deal with this kind of problem later; for now, let us realize that whenever we break up a set to count its elements more easily, it is very important that the resultant pieces be nonoverlapping.

Combining sets by taking their union and then counting the elements is relatively easy. When we try to combine sets by taking their Cartesian product, things become a bit more complicated. Suppose A is an $m-$set and B is an $n-$set.

RULE OF PRODUCT: If A is an $m-$set and B is an $n-$set, then $A \times B$ is an $mn-$set.

(Why doesn't it matter whether A and B overlap?) To see why this is so, think about how to form all of the ordered pairs in $A \times B$. There are m choices for the first element of each pair and n choices for the second element. If we select a particular element z of A and hold it fixed as the first element of the pair, we can form n different elements of $A \times B$ with z as the first element. Since we can do this for each of the m different elements of A, we can thus generate mn pairs for $A \times B$. In general, we have a similar rule for a product of several sets.

GENERALIZED RULE OF PRODUCT: Let $| A_i | = n_i$ for $i = 1, 2, ..., r$. Then the Cartesian product $A_1 \times A_2 \times ... \times A_r$ is an $(n_1 n_2 ... n_r)-$set. Let us see how these rules can help us to count things in typical computing situations.

SAMPLING

Suppose we have a set of cardinality n and we want to sample this set. This means that we want to pick an element at random from the set. For instance, suppose we are testing a program and we want to check the contents of the registers periodically. In simple terms, a register may contain any of n different values which we can consider as n distinct elements (for some n which depends on the size of the register). For instance, an eight$-$bit binary register can hold 256 different numbers. We want to look at this register r times during the running of the test program. We will end up with r "looks."

It is important to note two things. First, the r peeks into the register may not be distinct. That is, we may very well have at least two times when the register contains the same value. This will certainly be true if we look more times than there are elements in the set of possible values (more than

256 times, in the example above). Second, the ordering of the r "looks" is very important. We don't want to consider these r things as a set, because our set notation would collapse any multiple occurrences of an item and would ignore order. Instead, we want to consider this sequence as an ordered r—tuple. We can write the sample of register contents as

$$(x_1, x_2, x_3, ..., x_r)$$

where x_1 is the value of the first sample, x_2 is the value of the second, and so on. We call such a sample an r—**sample**, and we say that the sample is of **size** r. It is possible that $x_i = x_j$ for some i and j.

When are two samples equal? First, they must be of the same size. Second, since order is important, two r—samples $(x_1, x_2, ..., x_r)$ and $(y_1, y_2, ..., y_r)$ are equal if and only if $x_i = y_i$ for every i from 1 to r. We can count the number of possible r—samples of a set of size n using the generalized rule of product.

THEOREM **6.5** _____

The number of r—samples of an n—set is n^r.

_____

To prove this, let A be an n—set. We want to count all possible ordered r—tuples where each of the elements in the r—tuple can be chosen from an n—set. This means that we are choosing one element from A for the first position (i.e., we have n choices), one element from A for the second position (again, n choices), and so on, for r positions. In other words, we are forming all possible elements of the set $A \times A \times ... \times A$, where A is repeated r times. According to the generalized rule of product, we have the case where $A = A_i$ and $n = n_i$ for each i from 1 to r, so our set $A \times A \times ... \times A$ is an n^r—set. ■

_____

Such a result can be quite useful. For example, a byte has eight bits, so we can think of a byte as an eight-digit binary string. Our theorem tells us that there are 2^8 such binary strings, since we have two choices (0 or 1) for each digit.

We can count those strings which satisfy a certain condition or have a certain property. Let S be the set $\{ x \mid x$ is an 8—bit binary string $\}$. Then, $|S| = 2^8$. Let P be the set $\{ x \in S \mid x$ has an even number of 1s $\}$. We know that P is a proper subset of S, and we want to find the cardinality of

P. We can partition *S* into two sets, *P* (as defined above) and *Q*, where *Q* is the set of all elements of *S* having an odd number of ones. It is clear that *S* is the union of *P* and *Q* and that *P* and *Q* have no elements in common. Thus, by the rule of sum, $|S| = |P| + |Q|$.

Let us define a mapping $f: Q \rightarrow P$ where f is defined in the following way:

> For each byte $a_1 a_2 a_3 \ldots a_8$ in *Q*, a^* be the rightmost 1. Then, define $f(a_1 a_2 a_3 \ldots a_8)$ as the byte $b_1 b_2 \ldots b_8$ where the b_i are defined according to the following rules:
> For $i = 1, 2, \ldots, 6, b_i = a_i$.
> If $a_7 a_8 = 11$, then $b_7 b_8 = 10$.
> If $a_7 a_8 = 00$, then $b_7 b_8 = 01$.
> If $a_7 a_8 = 10$, then $b_7 b_8 = 00$.
> If $a_7 a_8 = 01$, then $b_7 b_8 = 11$.

It is left as an exercise to show that f is a bijection. Thus, it must be the case that $|P| = |Q|$, so $|P| = 2^7$.

Other applications of Theorem 6.5 abound in computing. For example, a functional circuit with r inputs, each of which has n possible values, has n^r possible input configurations. It is important to know this when testing the circuit to be sure that all possible configurations have been tested.

When writing a compiler, it is useful to know which combinations of expressions are equivalent in the language and which are not. For example, given a logical expression connected by AND connectives, there are some assignments of truth values which result in the expression's being TRUE and others which result in FALSE. The expression

A AND *B* AND *C*

is not true if any of *A*, *B*, or *C* is FALSE. Of course, expressions can be much more complex. The expression

(NOT *A* AND *B* AND NOT *C*) OR (*A* AND *C*)

is not as easy to evaluate. Expressions of this form resemble the POS form with which we dealt in Chapter 4. Here, we have groups of variables connected by AND operators which are in turn connected by OR operators. Such expressions are said to be in **conjunctive normal form**. Similarly, an SOP corresponds to an expression in **disjunctive normal form**, where groups of variables connected by OR connective are in turn connected by AND operators. How can the compiler check an expression in conjunctive normal form to find all possible variable combinations which make this expression true? One way is to try each combination, one by one. If there are n logical variables, then Theorem 6.5 tells us that there are 2^n combinations to evaluate, since each variable can be either 0 or 1 (so we have two choices for each variable). Unfortunately, those who analyze algorithms such as this one suggest that this is the best technique available for deciding which variable combinations are acceptable. It is for this reason that this sort of problem is called a 2^n **problem**.

PERMUTATIONS

Consider now the $r-$sample $(a_1, a_2, ..., a_r)$ from an $n-$set where order is still important but the a_i are distinct. In this case, the $r-$sample is called an $r-$**permutation of n elements**. What happens when we try to count the number of $r-$permutations of n elements? Let us look at a simple example. Suppose we have the n objects in a drawer, and we reach into the drawer to select one element. As in the previous situation, we have n possibilities for the element that we pull out. However, in the previous instance, we then return the element to the drawer before we choose again. In this instance, our elements have to be distinct, so we must put aside the chosen element and return to a drawer with only $n-1$ elements in it. Continuing in this manner, we choose one of the remaining $n-1$ elements, put it aside to be the second element of our $r-$tuple, and go back in the drawer to choose one of the remaining $n-2$ elements. Thus, it is easy to see that

THEOREM 6.6 _____

The number of $r-$permutations of n elements is

$$n(n-1)(n-2) ...(n - r + 1).$$

_____

Again, we can use the generalized rule of product. Let $A = A_i$ for each i, and let $n_1 = n$, $n_2 = n-1$, ..., and $n_r = n - r + 1$. ∎

We use the notation $P(n, r)$ to stand for the number of $r-$permutations of n elements.

When r and n are equal, we are counting the number of $n-$permutations of n elements. What does this mean? We have n elements, and we choose one of the n, set it aside, choose one from the remaining $n-1$, put it aside, choose one from the remaining $n-2$, and so on, until all of the original n elements are chosen. Because order is important, we are really lining up the elements as we choose them. Thus, $P(n, n)$ really counts the ways in which we can order n elements.

For example, suppose we want to find the number of possible five—letter words made up of the letters a, b, c, d, and e. This would be useful in decoding a message, for instance, or in playing a popular word game. Because the letters are distinct, we know they can be arranged in $P(n,n)$ = $n!$ = $5!$ = 120 ways. Now let us look at some more general questions. How many different five—letter words can we make out of the English alphabet of 6 letters? Allowing repetitions of letters, Theorem 6.4 tells us that the answer is 26^5 = $11,881,376$. However, if the letters must be distinct, then Theorem 6.5 tells us that the answer is $P(26, 5)$ = $26 \times 25 \times 24 \times 23 \times 22$ = $7,893,600$.

There is another way of writing $P(n, r)$ which may shed more light on how the counting process works. Note that we can write $n(n - 1)(n- 2)$ ...$(n - r + 1)$ as the equivalent form

$$\frac{n(n- 1)(n- 2) \, ...(n - r \, + 1)(n - r)(n - r - 1) \, ...(3)(2)(1)}{(n - r)(n - r- 1) \, ...(3)(2)(1)}$$

which is the same as writing

$$\frac{n \, !}{(n - r)!}$$

In other words, this says that

$$\frac{P(n, r) = P(n, n)}{P(n - r, n - r)}$$

Consider this formula in intuitive terms. We are arranging n items in a line (i.e., ordering n items, or choosing n of n items one at a time, keeping track of the order in which the items are chosen). We can mark n positions onto which we will place these items, and we can break the n positions into two groups: the first r positions and then the remaining $n - r$. First, we will put r of the n items in the first r positions; there are $P(n, r)$ ways of doing this. Then, we will put $n - r$ of the remaining $n - r$ items in the next set of positions; there are $P(n - r, n - r)$ ways of doing this. Using the rule of product, we have shown that

$$P(n, n) = P(n, r) \times P(n - r, n - r)$$

which is the same as the previous formula.

We can use the formula for the number of permutations in everyday computing situations. For example, suppose your computer assigns your output to one of five printers. Because each printer is at a different remote location, the order of assignment is important. We want to schedule no more than one program's output per printer. If there are three sets of output to be assigned, the number of ways that the system can do this is P(5, 3). This is because there are five choices for the first output, four for the second, and three for the third.

In another case, suppose that the computing center wants to give each user an identification string which is of the form *LLLDD*. Here the *L* signifies distinct letters and the *D* represents distinct digits. The number of different user identification strings is thus

P(26, 3) × P(10, 2)

CIRCULAR ARRANGEMENTS

Let us use a similar line of reasoning to examine the arrangements of items around a circle. The first thing to note is that an arrangement such as the one shown in Figure 6.5 is the same as the arrangements shown in Figure 6.6. This is because a given arrangement can be rotated around the circle; only the relative positions of the items to each other is important, so all of the rotated arrangements are equivalent to the original.

If we were to arrange the *n* items in a line, we would have P(*n*, *n*) arrangements. Wrapping each of those arrangements around the circle, we see that we are counting the rotations of each arrangement *n* times, which is *n* − 1 times more than we want to count it. So the number of distinct arrangements of *n* items in a circle is really

$$P(n, n) = \frac{(n-1)!}{n}$$

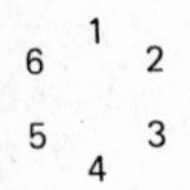

Figure 6.5

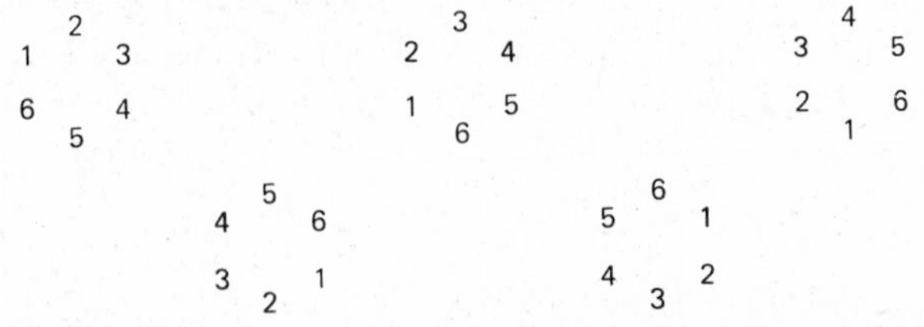

Figure 6.6

Another way of seeing this is to imagine that a first item is picked and placed in a particular fixed position. Then, arranging the remaining $n-1$ items around the circle is equivalent to lining up the remaining $n-1$ items. The number of ways of doing this is of course $P(n-1, n-1)$, or, in other words $(n-1)!$.

A simple example of this is to consider the determination of a bus route. Suppose there are 15 stops to be made on the bus route, and the bus route is to start and end at the same place. Then, this is an instance where we have a circular arrangement of the 15 bus stops, so there are 14! possible routes. However, if specific starting and ending places are named, then we must reconsider this problem. If the starting and ending places are different, there are 13! possible routes.

PERMUTATIONS OF NONDISTINCT OBJECTS

So far, we have considered arrangements of items that are either all distinct or all the same. It is more often the case that some of the items can somehow be distinguished from the other items. The simplest case is where the items can be split into two groups. For instance, in decoding a message, the message may consist of letters and numbers. We may want to know how many arrangements there are of seven items, two of which are numbers and five of which are letters. Let us consider how to enumerate permutations of this kind where not all of the objects are distinct but where the groupings of them matter.

Suppose that we have n objects, but not all of the objects are distinct. We have k_1 objects of the first kind, k_2 objects of the second kind, k_3 objects of the third kind, and so on, until we finally have k_m objects of the mth kind. Suppose further that we can somehow mark these objects so that objects of any one kind can be separated one from the other within that kind. For instance, if we have a group of the same letter, we can assign numbers to the letters so that they become letter-1, letter-2, letter-3, and so on. Now we have n distinct objects, so there are $n!$ ways of permuting these objects. However, the markings to distinguish all of the objects are just temporary. Once the markings are removed, any letter is equivalent to any other letter. We want to make sure that we don't count multiple times the arrangements that will look the same once the distinguishing markings are removed. There are $k_1!$ permutations of the objects in the first group, $k_2!$ permutations of the objects in the second group, and in general $k_i!$ permutations of the objects in the ith group. Thus, to avoid overcounting, we must divide $n!$ by each of the $k_i!$. What we have shown, then, is the following:

THEOREM 6.7 _____

Given n objects, where there are k_i objects of type i, $i = 1, 2, ..., m$, and $k_1 + k_2 + ... + k_m = n$, the number of permutations of these n objects is

$$\frac{n\,!}{k_1\,!\;k_2\,!\;...\;k_m\,!}$$

Let us look at some examples to see how this formula is applied. Morse code is made up of dots and dashes. We want to know in how many ways we can arrange six dots and four dashes. The dots make up the group of items of the first kind, and the dashes make up the group of items of the second kind. There are 10! ways of arranging ten things, 6! ways of arranging the dots, and 4! ways of arranging the dashes. Thus, there are $10!/(6! \times 4!) = 210$ ways of arranging six dots and four dashes to form an arrangement of ten.

As another example, we turn to computing. A WARTHOG 660 is a front—end processor to five mainframe computers at a local university. There are 32 incoming phone lines to the WARTHOG. In how many ways can the front—end processor assign lines to computers so that eight will go to computer 1, four to computer 2, four to computer 3, ten to computer 4, and six to computer 5? From the formula derived above, the answer must be

$$\frac{32!}{8!\;4!\;4!\;10!\;6!}$$

UNORDERED SELECTIONS

So far, we have been considering situations in which the order of our aligning or arranging or selecting has been important. Let us now consider times where the order is not significant. For instance, we are choosing students for a special experimental class. We may be choosing r students from a possible n students, but it makes no difference if a student is chosen first or last; all that matters is whether or not a student is chosen at all. In other instances, we may choose something or someone more than once. For example, the university may want to award 12 scholarships to the student body. Each scholarship has a different purpose, and it is possible that one student can win multiple scholarships.

Let us formalize our ideas by considering selections from a set S. If $\{ a_1, a_2, ..., a_r \}$ is an unordered collection of r elements of S, it is possible that these elements are not distinct. In other words, we may have $a_i = a_j$ for at least one pair i and j. With each distinct element, then, we can associate a **multiplicity** which counts the number of occurrences of the element in this selection. Thus, we say that two collections

$$\{\ a_1, a_2, ..., a_r\ \}$$

and

$$\{\ b_1, b_2, ..., b_r\ \}$$

are the same if the elements and corresponding multiplicities are equal. We refer to the collection

$$\{\ a_1, a_2, ..., a_r\ \}$$

as an **unordered selection of size** r, and some books call this an $r-$**selection of the set** S. Note that if each element of an $r-$selection of S is of multiplicity 1 (that is, there are no repeats in the collection), then the $r-$selection is also an $r-$subset of S. In this special case, we call our collection an $r-$**combination of** n **elements**. How can we count them? The next theorem tells us.

THEOREM 6.8 _____

The number of $r-$subsets of an $n-$set is

$$\frac{P(n, r)}{-r\ !}$$

_____

We know that we can find $P(n, r)$ $r-$permutations of our $n-$set. The only difference between a permutation and an $r-$subset is that order was important when we generated each permutation. Since each $r-$permutation can be ordered in $r!$ ways, this means that we are counting each collection $r!$ times when we disregard order. Thus, we have

$$\frac{P(n, r)}{r\ !}$$

different $r-$subsets. ∎

Often, you will see this number written in different ways. We define

$$C(n, r) = \frac{P(n, r)}{r!}$$

Another way of writing $C(n, r)$ is

$$\binom{n}{r}$$

This is often read as *n choose r* to indicate that it is the number of ways of choosing r items from n items. Viewing $C(n, r)$ in this way, it is clear from the meaning of $C(n, r)$ that if r is larger than n, then $C(n, r)$ is 0. This is a mathematical way of saying that there is no way to choose r things from n things when there aren't even enough things to choose. In particular, for $r > 0$, we know that

$$C(0, r) = 0$$

From the definition of 0! as 1, it follows that

$$C(n, 0) = 1$$

for $n \geq 0$. In this way, $C(n, r)$ is defined for all nonnegative integers n and r.

Let us see how to use the numbers $C(n, r)$. Suppose the systems programmer wants to schedule backups three times a week. The number of ways to do this would be $C(7, 3)$ if the computer is running seven days a week, or $C(5, 3)$ if the work week is only five days long.

In another instance, suppose you need to count the number of bytes containing exactly 2 zeros. You can think of this in terms of filling each of the eight bits of a byte with either a 0 or a 1 in each bit. You can choose two of the eight bits to contain the 0s, and the remaining six bits must thus contain 1s. There are $C(8, 2) = 28$ ways of choosing the two bits, so there must be 28 bytes with exactly two 0s.

Finally, let us look at the program SYSPERF which monitors the performance of your computer system. As each user logs on, the user is assigned by the operating system to one of 32 areas of memory known as a system partition. SYSPERF samples data from 10 of the 32 system partitions. There are $C(32, 10)$ ways of doing this. If partition X is always to be included in the sample, then there are C(31, 9) ways of choosing the other partitions in the sample. If X is always to be excluded, however, then there are C(31, 10) ways of choosing the partitions. In how many ways can we choose the partitions so that at least one of partitions X and Y is included? Including both means that there are C(30, 8) choices. Including X but not Y yields C(30, 9) choices, as does including Y but not X. Thus the total number of choices of partitions which include at least one of X and Y must be C(30, 8) + 2C(30, 9).

SELECTION WITH REPETITION

In dealing with $r-$subsets, we are not allowing for any repetitions. Suppose we turn now to $r-$selections, where repetitions are allowed. In this case, we have the following:

THEOREM **6.9** _____

The number of $r-$selections of an $n-$set is

$$C(n + r - 1, n - 1) = C(n + r - 1, r)$$

_____

Without loss of generality, let the set $J = \{\ 1, 2, ..., n\ \}$ of positive integers less than or equal to n be our $n-$set. Every $r-$selection from this set can be written as $\{\ i_1, i_2, ..., i_r\ \}$, where $i_1 \leq i_2 \leq ... \leq i_r$. Consider next the set of positive integers up to and including $n + r - 1$. Call this set J'. Then, the set

$$\{\ i_1 + 0, i_2 + 1, ..., i_r + r - 1\ \}$$

is an $r-$subset of J'. It is clear that we can define a bijection from J to J ' by mapping every element of the form

$$\{\ i_1, i_2, ..., i_r\ \}$$

to an element of the form

$$\{\ i_1 + 0, i_2 + 1, ..., i_r + r - 1\ \}$$

Theorem 6.8 tells us that the cardinality of J' is C(n + r - 1, r), so we are done. ■

ORDERED PARTITIONS

In Chapter 5, we examined the lattice of SP partitions in order to decompose a finite state machine. In that lattice, we compared one partition to another. Here, we examine partitions in a different light by ordering the elements involved in the partition. For example, suppose we want to partition the senior class of 100 students into five committees of 7, 12, 35, 21 and 15 students, respectively. In how many ways can we choose the committees so that each student serves on exactly one committee?

Remember that when we partition a set A of cardinality n, we find disjoint subsets B_1, B_2, ..., B_k, where

$$A = \bigcup_{i=1}^{k} B$$

and $B_i \cap B_j = \phi$ for every pair of distinct i and j. Suppose the cardinality of each B_i is b_i . This means that each B_i is a b_i-subset of A. Then,

$$b_1 + b_2 + ... + b_k = n$$

Sometimes, $\{ B_1, B_2, ..., B_k \}$ is called a $(b_1, b_2, ..., b_k)$**-partition of** A. If we assign an order to the elements involved, we have an **ordered** $(b_1, b_2, ..., b_k)$**-partition of** A.

THEOREM 6.10_____

The number of ordered $(b_1, b_2, ..., b_k)$—partitions of an n—set is

$$\frac{n!}{b_1 ! \; b_2 ! \; b_3 ! \; . \; . \; . \; b_k !}$$

_____

Theorem 6.8 tells us that we can select the first b_1 items from n items in $C(n, b_1)$ ways. We then have $n - b_1$ items remaining, and we want to choose b_2 of those; we can do this in $C(n - b_1, b_2)$ ways. We continue in this fashion until we have $n - b_1 - b_2 - ... - b_{k-1}$ items left from which to choose b_k. We make our final choice in $C(n - b_1 - b_2 - ... - b_{k-1}, b_k)$ ways. The rule of product says that the number of ways of determining this ordered $(b_1, b_2, ..., b_k)$-partition is thus

$$C(n, b_1) \times C(n - b_1, b_2) \times ... \times C(n - b_1 - b_2 - ... - b_{k-1}, b_k$$

If we multiply this out (this is left for you as an exercise), we have our result. ■

BINOMIAL COEFFICIENTS

There are many relationships among the numbers $C(n, r)$ which are useful when using counting arguments to prove theorems. For example, from the definition of $C(n, r)$, it is left as an exercise to show that

$$C(n, r) = C(n-1, r) + C(n-1, r-1)$$

Intuitively, this equation says that if you are choosing r items from a set of n items, you can designate one item as a specially marked item. The $r-$set you then choose either includes this marked item or it doesn't. If it doesn't, then there are $n-1$ items from which to choose your r items, and there are $C(n-1, r)$ ways to do that. However, if you *do* include the marked item, then you have to choose $r-1$ other items from the remaining $n-1$ items. There are $C(n-1, r-1)$ ways to do that. Hence, the result.

This formula is what is behind the convenient method you have probably used for generating binomial coefficients: you form a triangle of them, adding the two nearest in the row above to generate the next position (Figure 6.7). This is known as **Pascal's triangle**. You used this triangle in algebra to generate coefficients for the equation

$$(x + y)^n = C(n, 0)x^n + C(n, 1)x^{n-1} y + ... + C(n, n-1)xy^{n-1} + C(n,n)y^n$$

which is the **binomial theorem**.

```
                1   1
              1   2   1
            1   3   3   1
          1   4   6   4   1
        1   5  10  10   5   1
      1   6  15  20  15   6   1
    1   7  21  35  35  21   7   1
  1   8  28  56  70  56  28   8   1
```

Figure 6.7

The theorem has a useful interpretation in terms of what we have presented in this chapter. Suppose we have a set of n symbols S_i, and each symbol is of the form

$$(x + y) = S_i$$

for i ranging from 1 through n. We can distinguish these symbols temporarily (just as we did the letters LLL in a previous example) by calling them

$$(x + y)_1, (x + y)_2, ..., (x + y)_n$$

Then, expanding the expression $(x + y)^n$ means determining coefficients for each of the terms of the form $x^{n-r} y^r$. However, the coefficient for this term is just the number of r−subsets of our set of n specially marked symbols. Because Theorem 6.8 tells us that this number must be $C(n, r)$, we have thus proved the binomial theorem.

We can use the binomial theorem to generate several useful identities which may help us in later chapters to count cases or items. First, note that we can write a particular case of the binomial theorem where Y is equal to 1:

$$(1 + x)^n = \sum_{k=0}^{n} C(n, k)\, x^k$$

This equation in turn generates several important theorems.

THEOREM 6.11 _____

$$\sum_{k=0}^{n} C(n, k) = 2^n$$

To prove this theorem, let $x = 1$ in the expansion of $(1 + x)^n$. ∎

Combinatorially, we can interpret Theorem 6.11 as telling us that, in counting the number of subsets of an n−set, we can count all of the 0−sets, all of the 1−sets, then all of the 2−sets, then all of the 3−sets, and so on, until we have counted all of the sets. This is because $C(n, 0)$ counts the empty set, $C(n, 1)$ counts all of the 1− subsets, $C(n, 2)$ counts all of the 2−subsets, and, in general, $C(n, k)$ counts all of the k−subsets of the n−set. We know from Chapter 1 that there are 2^n subsets, so we have derived our equation in yet another way.

THEOREM 6.12_____

$$\sum_{k=0}^{n}(-1)^k \, C(n, \, k) = 0$$

_____

To prove this, let x be -1 in the expansion of $(1 + x)^n$. ▮

_____

This result reflects the fact that the binomial coefficients, when lined up as Pascal's triangle, are symmetric on each row. In other words, this result follows from the fact that the signs of the terms alternate and

$C(n, \, k) = C(n, \, n - k).$

Similar results can be obtained by using similar methods. For instance, we can prove

$$\sum_{k=0}^{m}C(r, \, k) \, C(s, \, m - k) = C(r + s, \, m)$$

by using the fact that

$$(1 + x)^r \, (1 + x)^s \, (1 + x)^{r + s}$$

Alternatively, we can interpret this combinatorially by considering the number of ways of choosing a committee of m people from a group of r men and s women. The summation tells us that we can choose k of them from the r men (hence the term $C(r, \, k)$, and the remaining $m - k$ from the set of s women to end up with a set of m people chosen from a total of $r + s$ people. Either way of deriving the result is acceptable, and both the algebraic method and the combinatorial method are used throughout computer science to interpret results or to generate new ones.

THE PRINCIPLE OF INCLUSION/EXCLUSION

When counting the elements of a set, we have seen that it is useful to break the set up into pieces that can be counted more easily; then, we reconstruct the original set by putting the pieces back together in some fashion. Even the rules of sum and product make use of this trick by looking at a set in terms of union or Cartesian product. In this section, we return to the idea of uniting pieces of a set in order to count the entire set. Recall that the cardinality of the union of two sets is equal to the sum of the cardinalities of the two sets only when the two sets are disjoint. This means that when the two sets overlap, we are double—counting the elements that are in the intersection of the two sets. More formally, we write this as

$$| A_1 \cup A_2 | = | A_1 | + | A_2 | - | A_1 \cap A_2 |$$

For instance, suppose we are counting the number of people in class with brown hair and green eyes. If A_1 is the set of people in class with brown hair and A_2 the set of people in class with green eyes, then by simply adding $| A_1 |$ and $| A_2 |$, we count twice all those people whose hair is brown and who also have green eyes. In other words, we are counting twice the people in $| A_1 \cap A_2 |$.

What happens when we have three sets? Let's look at the Venn diagram in Figure 6.8. We double—count the elements in the intersection of any pair of sets. If we subtract the number of elements in the intersections of pairs of sets, we eliminate the extra count of those elements appearing in two of the three sets, but we would subtract one too many times any element which is in the intersection of all three sets. Thus, we want to add back in any element which is in the intersection of all three of the sets. Our final formula is therefore

$$
\begin{aligned}
| A_1 \cup A_2 \cup A_3 | = & \, | A_1 | + | A_2 | + | A_3 | \\
& - | A_1 \cap A_2 | - | A_1 \cap A_3 | - | A_2 \cap A_3 | \\
& + | A_1 \cap A_2 \cap A_3 |
\end{aligned}
$$

For four sets, the situation is a bit more complex, but the counting principle remains the same. Define the sets A_1, A_2, A_3, and A_4 as follows:

$A_1 = \{$ all people with brown hair $\}$
$A_2 = \{$ all people with blue eyes $\}$
$A_3 = \{$ all people with false teeth $\}$
$A_4 = \{$ all people with mustaches $\}$

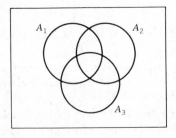

Figure 6.8

To determine the number of people in the union of these four sets, we keep in mind the Venn diagram of Figure 6.9. First, we add all of the elements in each set, so we have as a first term

$$\sum_{i=1}^{n} |A_i|$$

where n is equal to 4. As before, we are overcounting, and the same arguments as above mandate that we subtract the sum of intersection of pairs and add back the sum of the intersection of triples. What happens to an element which is in all four of the sets? Someone who has brown hair, blue eyes, false teeth, and a mustache is counted four times in the term

$$\sum_{i=1}^{n} |A_i|$$

The next term,

$$\sum_{1 \le i < j \le n} |A_i \cap A_j|$$

has our brown-haired, blue-eyed, false-toothed, mustachioed person in every set of the summation. Thus, our person appears six times in this term (why?), so we have subtracted too many. Likewise, adding back triples, we have four triples of sets (why?), so we have counted our person $4 - 6 + 4 = 2$ times. Thus, we must add another term to subtract all of the elements in the intersection of all four sets.

We can continue to increase the number of sets in a similar fashion, and the general result is known as the

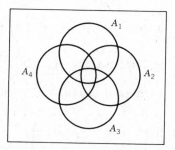

Figure 6.9

PRINCIPLE OF INCLUSION/EXCLUSION

For sets A_i, $i = 1, 2, ..., n$,

$$\left| \bigcup_{i=1}^{n} A_i \right| = \sum_{i=1}^{n} |A_i| - \sum_{1 \le i < j \le n} |A_i \cap A_j| + \sum_{1 \le i < j < k \le n} |A_i \cap A_j \cap A_k|$$

$$+ ... + (-1)^{n-1} \left| \bigcap_{i=1}^{n} A_i \right|$$

To prove this, we use induction on k, the number of sets.
Basis: The case where k is 1 is trivial, and the cases for $k = 2$ and $k = 3$ have been shown above.
Induction Step: Suppose we know that our result is true for all integers k such that k is less than t, and we want to show that the result is true for $k = t$. Consider the union of all t sets to be the union of two sets, namely, the set A_t with the set $A_1 \cup A_2 \cup ... \cup A_{t-1}$. This means that

$$| A_1 \cup A_2 \cup ... \cup A_t | = | A_1 \cup A_2 \cup ... \cup A_{t-1} | + | A_t | - | A_t \cap (A_1 \cup A_2 \cup ... \cup A_{t-1}) |$$

However, we can rewrite $A_t \cap (A_1 \cup A_2 \cup ... \cup A_{t-1})$ as

$$(A_t \cap A_1) \cup (A_t \cap A_2) \cup ... \cup (A_t \cap A_{(t-1)})$$

and, using the inductive hypothesis for the $t - 1$ sets $A_t \cap A_i$, we can write

$$|(A_t \cap A_1) \cup (A_t \cap A_2) \cup ... \cup (A_t \cap A_{(t-1)})| =$$
$$|A_t \cap A_1| + |A_t \cap A_2| + ... + | A_t \cap A_{t-1}|$$
$$- |(A_t \cap A_1) \cap (A_t \cap A_2)| - |(A_t \cap A_1) \cap (A_t \cap A_3)| - ...$$
$$+ |(A_t \cap A_1) \cap (A_t \cap A_2) \cap (A_t \cap A_3)| + ...$$
$$- ...$$
$$+ (-1)^{t-2} |(A_t \cap A_1) \cap (A_t \cap A_2) \cap ... \cap (A_t \cap A_{(t-2)})|$$

Condensing the long intersections, we get

$$|(A_t \cap A_1) \cup (A_t \cap A_2) \cup ... \cup (A_t \cap A_{(t-1)})| =$$
$$|A_t \cap A_1| + |A_t \cap A_2| + ... + | A_t \cap A_{t-1}|$$
$$- |A_t \cap A_1 \cap A_2| - |A_t \cap A_1 \cap A_3| - ...$$
$$+ |A_t \cap A_1 \cap A_2 \cap A_3)| + ...$$
$$- ...$$
$$+ (-1)^{t-2} |A_t \cap A_1 \cap A_2) \cap ... \cap A_{(t-1)}|$$

We also know that for the sets A_1 through A_{t-1}, the inductive hypothesis holds. This means that

$$|A_1 \cup A_2 \cup ... \cup A_{t-1}| =$$
$$|A_1| + |A_2| + ... + |A_{t-1}|$$
$$- |A_1 \cap A_2| - |A_1 \cap A_3| - ...$$
$$+ ...$$
$$+ (-1)^{t-1} |A_1 \cap A_2 \cap ... \cap A_{t-1}|$$

By substituting this formula and the previous one into the formula for the cardinality of the union of the sets $A_1, A_2, ..., A_t$, we have our result. ■

The principle of inclusion/exclusion looks complicated, but it is the logical extension of a fairly simple concept. It has wide applications and can lead to powerful results. For example, suppose you are trying to evaluate the performance of your computer. There are many programs available to run on the machine and you want to count up those that do not burden the system. Suppose there are 500 programs resident and ready to run. Of these, 295 link to the system library, 300 require communication with a remote site, and 175 access data on the auxiliary disk. There are 200 that link to the system library and require communication to a remote site, 125 that link to the system library and access data on the auxiliary disk, and 75 that communicate with a remote site and access the auxiliary disk. Finally, 60 of the programs link to the system library, communicate with a remote site, and communicate with the auxiliary disk. How many of the programs on the machine require neither the system library, the auxiliary disk, nor communication with a remote site? To answer this, let us define the set A_1 to be the set of all programs that link to the system library, A_2 the set of all programs that communicate with a remote site, and A_3 the set of all programs that access data on an auxiliary disk. Then we have

$$| A_1 | = 295$$
$$| A_2 | = 300$$
$$| A_3 | = 175$$
$$| A_1 \cap A_2 | = 200$$
$$| A_1 \cap A_3 | = 125$$
$$| A_2 \cap A_3 | = 75$$
$$| A_1 \cap A_2 \cap A_3 | = 60$$

This is pictured in Figure 6.10. We want to calculate the quantity $| A_1 \cap A_2 \cap A_3 |$, but by De Morgan's Laws, we know that

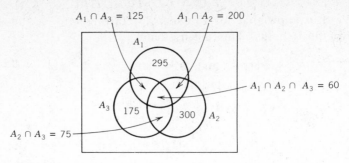

Figure 6.10

$$\overline{A_1} \cap \overline{A_2} \cap \overline{A_3} = \overline{(A_1 \cup A_2 \cup A_3)}$$

Thus, we have

$$|\overline{A_1 \cup A_2 \cup A_3}| = 500 - |A_1 \cup A_2 \cup A_3|$$

The principle of inclusion/exclusion tells us how to calculate the term on the right:

$$
\begin{aligned}
|A_1 \cup A_2 \cup A_3| &= |A_1| + |A_2| + |A_3| - |A_1 \cap A_2| - |A_1 \cap A_3| \\
&\quad - |A_2 \cap A_3| + |A_1 \cap A_2 \cap A_3| \\
&= 295 + 300 + 175 - 200 - 125 - 75 + 60 \\
&= 430
\end{aligned}
$$

This means that there are $500 - 430 = 70$ programs that do not link to the system library, do not communicate to a remote site, and do not access the auxiliary disk.

For another example, consider a computer that encodes its data in strings in the following way. Its words are comprised of eight digits, where each digit is chosen from the set $\{ 0, 1, 2, 3, 4, 5, 6, 7 \}$. Any word which contains at least one 0, one 1 and one 2 is handled in a special way by the computer's operating system. We know that there are 8^8 possible words of this type. Let A_0 be the set of bytes with no occurrences of 0, A_1 be the set of bytes with no occurrences of 1, and A_2 the set of bytes with no occurrences of 2. Then, $A_0 \cup A_1 \cup A_2$ is the set of bytes in which one or more of the digits 0, 1 and 2 is missing. We have

$$|A_0| = |A_1| = |A_2| = 7^8$$

We also know that

$$|A_0 \cap A_1| = |A_0 \cap A_2| = |A_1 \cap A_2| = 6^8$$
$$|A_0 \cap A_1 \cap A_2| = 5^8$$

so

$$|A_0 \cup A_1 \cup A_2| = 3(7^8) - 3(6^8) + 5^8$$

by the principle of inclusion/exclusion. Therefore, the number of bytes containing at least one 0, one 1, and one 2 must be

$$8^8 - [3(7^8) - 3(6^8) + 5^8].$$

RECURRENCE RELATIONS

Often, it is easy to define a function in terms of previously computed values of the function. For example, we compute n ! as the product of the positive integers less than or equal to n. However, if we were writing a program to generate and print out all of the factorials from 1 to 1000, we would not want to use that definition to calculate each factorial because we would continually multiply many of the same numbers.

Instead, we would define factorials in the following way:

$1! = 1$
$n ! = n \times ((n-1)!)$ for $n \geq 2$

In using this definition in a program, the only calculation involved for the next factorial is the multiplication of the previous factorial by the next integer.

Let us examine a similar situation. Let A be an $n-$set and a an element of A. For r less than n, some $r-$permutations of elements of A will contain a; the rest will not. Those containing a will have a in one of the r positions, and the remaining $r-1$ positions will be chosen from the $n-1$ elements of the set $A - \{ a \}$. Thus, there are $P(n-1, r-1)$ ways of assigning the remaining positions. Because there are r ways of choosing the element a in A, and because there are $P(n-1, r)$ ways of choosing r elements from $A - \{ a \}$, we can conclude that

$$P(n, r) = r \; P(n-1, r-1) + P(n-1, r)$$

This formula generates the next $r-$permutation of an $n-$set by using values from previous generations. In order to calculate the numbers $P(n, r)$, though, we need one more piece of information. We need a starting value. The starting value, known as a **boundary condition**, acts as a seed to begin the generation of the rest of the values. Sometimes, several boundary conditions are required to begin the calculations.

This kind of definition is known as a **recurrence relation**. As you can see, the number to be calculated occurs in some form on both sides of the equality. In a sense, then, unless you already know some of the values to be calculated, you cannot calculate the others. Of course, it is unacceptable to define one level of a recurrence relation in terms of itself. For example,

$Q(n, r) = 2 Q(n, r)$

is impossible to compute.

In contrast to a recurrence relation is the **closed form solution**. In this case, you don't need to know any other values. The closed form solution is a formula by which any term in the recurrence may be calculated by itself, independent of the other values of the recurrence. In the case of $P(n, r)$, Theorem 6.6 gives us a closed form solution. We must know n and r, but then we have enough information to find $P(n, r)$ quickly. Knowing n and r does not help us much in the recurrence relation for $P(n, r)$, however. A general rule of thumb is that recurrence relations are useful when it is necessary to generate a large range of values, but the closed form solution is better for the calculation of a single instance.

Recurrence relations can generate other equations. For example, recall that

$C(n, r) = C(n-1, r-1) + C(n-1, r)$

By expanding the first term repeatedly, we have

$C(n, r) = C(n-1, r) + C(n-2, r-1) + ... + C(n-1-r, 0)$

Then, by expanding the last term, we have

$C(n, r) = C(n-1, r-1) + C(n-2, r-1) + ... + C(r-1, r-1)$

This final equation is the rule that generates Pascal's triangle.

Consider now a geometric example. Let $P(n)$ be the maximum number of regions into which a plane is divided by n straight lines. We define $P(0)$

= 1. Then, a recurrence relation for P(n) can be written as

$$P(n) = P(n-1) + n$$

What if we want the closed form solution? It is left as an exercise to show by induction that

$$P(n) = \frac{n(n+1)}{2} + 1$$

Note that this formula for P(n) does not require the knowledge of P(i) for all i less than n.

An important recurrence relation can be generated by looking at the set $S = \{ 0, 1 \}$ and the set A of n−samples of S. Let f (n) be the number of such samples that do not contain two successive zeros. As a boundary condition, let f (0) be 1. It is easy to see that f (1) = 2. For n at least 2, there are f(n−1) samples with the first component equal to 1 and f(n−2) with the first component equal to 0. Thus, we have the recurrence

$$f(n) = f(n-1) + f(n-2)$$

These numbers are known as the **Fibonacci numbers**. By beginning with the boundary conditions and repeatedly using the recurrence formula, we see that the series looks like

1, 2, 3, 5, 8, 13, 21, 34, ...

where the next number is the sum of the previous two.

This sequence shows up in astounding places. For instance, a nautilus shell spirals in a pattern related to the Fibonacci numbers, and the offspring of rabbits multiply according to this recurrence. One surprising mathematical result is that every integer can be written as a linear combination of the Fibonacci numbers. This means that for each integer k, we can write k as

$$k = a_0 \times f(0) + a_1 \times f(1) + a_2 \times f(2) + ... + a_i \times f(i) + ...$$

where the a_i are integers equal to 0 or 1.

The closed form solution for the Fibonacci numbers is

$$f(n) = \sum_{k=0}^{m} C(n-k+1, k)$$

where m is the largest integer contained in the quantity $(n + 1)/2$. How can we show that this formula is the same as the recurrence written above? One way is to let

$$g(n) = \sum_{k=0}^{m} C(n-k+1,k)$$

with m as stated. We can show that $g(0) = 1$, $g(1) = 2$, and that

$$g(n) = g(n-1) + g(n-2)$$

The proof is left as an exercise. It is easy to see that the numbers $g(n)$ must thus be the same as the numbers $f(n)$. This general technique is useful for demonstrating that a closed form solution is the same as the one generated by the recurrence relation.

Recall that in a previous example we looked at counting the ways of changing a computer password so that no character in the new password was in the same position as it was in the original password. Such a mixing of characters is called a **derangement**. If n is a positive integer and we take the sequence $< 1, 2, ..., n >$ and rearrange it into $< a_1, a_2, ..., a_n >$, where no a_i is equal to i for any i, then we have a derangement of the integers 1 through n. How can we count these derangements? If D_n is the number of derangements of the integers 1 through n, then it is easy to see that D_1 must be 0. For n larger than 1, we see that the first position can be anything except 1. Thus, we have $n-1$ choices for the first element of the derangement. Suppose the element in this first position is chosen to be k. Then, either a_k is 1 or it is not. If a_k is 1, then the rest of the derangement is the number of derangements of $n-2$ elements. If a_k is not 1, then we are rearranging the elements $1, 2, ..., k-1, k+1, ..., n$ in positions 2 through n, with 1 not in the kth position and every other element out of position. However, this is the same as permuting $n-1$ elements labeled 2 through n with all the elements displaced; this number is just D_{n-1}. What we have shown is a recurrence relation to generate the number of derangements of n elements:

$$D_n = (n-1)(D_{n-1} + D_{n-2})$$

The closed form solution is much more complex:

$$D_n = n! \left(1 - \frac{1}{1!} + \frac{1}{2!} - ... + (-1)^n \frac{1}{n!}\right)$$

It is left as an exercise to show that the closed form solution generates the same sequence of numbers as the recurrence relation.

As a final example, we turn to the analysis of algorithms. Suppose we want to find both the maximum and the minimum of a set of n elements. To simplify this discussion, let us assume that n is always a power of 2. Aho, Hopcroft, and Ullman, in *Design and Analysis of Computer Algorithms* (Addison-Wesley, 1974), present a procedure which makes use of the "divide and conquer" approach: divide the set of n elements repeatedly into two smaller sets, find the maximum and minimum for each half, and compare. The procedure can be described as follows:

```
procedure MAXMIN(S)
if |S| = 2 then
      begin
            let S = { a,b }
                return (MAX(a,b), MIN(a,b))
      end
else
      begin
            divide S into two subset S1 and S2,
              each with half of the elements;
            (max1, min1) < − −MAXMIN(S1);
            (max2, min2) < − −MAXMIN(S2);
            return(MAX(max1, max2),MIN(min1, min2))
      end
```

Note that the procedure is invoked within the procedure definition. Thus, this definition is a type of recurrence relation, since the procedure recurs within its own definition. You may have encountered this type of procedure before. It is known as a **recursive procedure**.

Algorithms such as this are analyzed to determine which ones are the fastest. In general, the greater the number of instructions, the slower the algorithm. In particular, the number of comparisons necessary to process an input of size n is an indicator of the speed of the algorithm. How can we find out the number of comparisons needed by a procedure or algorithm? We can use recurrence relations. Let $T(n)$ be the number of comparisons between elements of S required by the procedure MAXMIN to find the maximum and minimum elements of S. It is easy to see that $T(2)$ is 1, and it is left as an exercise to show that we can define the numbers $T(n)$ by a recurrence relation as well:

$$T(n) = 1 \text{ for } n = 2$$
$$T(n) = 2T(n/2) \text{ for } n > 2$$

The closed form solution to this recurrence is

$$T(n) = 3 \, (n \, / 2) - 2.$$

There are many other algorithms that have representations as recurrence relations. These include algorithms for various sort and merge techniques. As with the procedure shown above, not only is the procedure a recurrence relation, but so, too, can we express the number of comparisons as a recurrence relation. In your study and analysis of algorithms, you will see how useful these recurrence relations can be.

Much of what we addressed in this chapter deals with finite numbers. In the next chapter, we look at finite sets from a more geometric perspective: we draw the finite sets as graphs, and we examine the relationships among the elements.

EXERCISES

1. a. Prove that every real number has a unique binary representation.

b. Determine the binary representation for π. Show that this representation is nonrepeating.

2. In the section on permutations, a function f is described which changes the rightmost 1 of a byte to a 0. Show that this function is a bijection between the set of bytes with an even number of 1s and the set of bytes with an odd number of 1s.

3. In considering combinations of elements, we introduced an example of a program named SYSPERF which samples partitions to determine system performance. Solve this problem using the application of the principle of inclusion/exclusion.

4. The Warthog Computer Corporation manufactures computers with three options: extra memory, RAM disk, and a special "fail—safe" operating system. Last year, Warthog sold 200 computers. Among these, 45 computers had extra memory, 32 had RAM disk, and 24 had the special operating system. Also, there were six systems sold which had all of these options. Use this information to determine a lower bound for the number of Warthog computers sold last year with no options at all.

5. Let T be the set of $n-$digit ternary strings. This means that each element of T has n digits, each of which is either 0, 1, or 2. Calculate the number of these strings that have an even number of zeros.

6. You are writing an editor program for your new microcomputer. The basic editor menu at the entry level allows the user to choose among six options: adjust, insert, delete, copy, move, or quit. For each option selected, there is a menu for a second level of options: five for adjust, three for insert, four for delete, two for copy, three for move, and two for quit. Considering both levels, you want to write a subroutine for each option. How many subroutines must you write?

7. You are configuring your new computer system. The configuration menu looks like this:

a. Default input device
  Floppy disk drive 1
  Floppy disk drive 2
  Card reader
  Tape drive
  Console

b. Processor
  WARTHOG 4000
  WARTHOG 4400

c. Default output device
  Card punch
  Line printer
  Console
  Tape drive
  Floppy disk drive 1
  Floppy disk drive 2
  Winchester disk drive

How many possible different configurations are there?

8. A palindrome is a word that reads the same backward as it does forward. How many different eight−letter palindromes are there on the English alphabet?

9. a. In how many ways can two numbers be chosen from the set { 1, 2, 3, 4, 5, 6, 7, 8, 9, 10 } so that their sum is an even number? So that their sum is an odd number?

b. Show that $C(2n, 2) = 2C(n, 2) + n^2$. How does this formula apply to part a?

10. Prove that

$$\sum_{k=1}^{n} k(-1)^k \, C(n, k) = 0$$

for n greater than or equal to 1 by

a. Using an argument which expands the summation and collects terms.

b. Using a combinatorial (counting) argument.

11. Prove that

$$\sum_{k=r}^{n} (-1)^k \, C(k, \, r) \, C(n, \, k) = 0$$

for n greater than or equal to r by

a. Using an argument which expands the summation and collects terms.

b. Using a combinatorial (counting) argument.

12. Five distinct integers are chosen from the set of integers { 1, 2, ..., 100 }.

a. In how many ways can these integers be chosen?

b. For how many of these choices is the sum of the integers divisible by 5?

13. In the section on recurrence relations, the following closed form solution for the number $P(n)$ was given.

$$P(n) = \frac{n(n + 1)}{2} + 1$$

Prove the given equation by induction. Can you find another way to generate this equation? Derive the closed form solution from the recurrence relation.

14. Define the set S to be all of the positive integers less than or equal to the integer n. Let P be the set of pairs of elements of S whose sum is greater than n. What is the cardinality of P?

15. In how many ways can the letters a, b, c, d, e, and f be arranged so that b is to the left of c and c is to the left of d?

16. In how many ways can the letters a, b, c, d, e, and f be arranged so that the triples abc and def never occur?

17. PROGRAMMING PROBLEM: You are giving a dinner party for n married couples. You want to determine all possible seating arrangements around a circular table where no husband is sitting next to his wife. This is called the "probleme des menages." Write a program which accepts as input n pairs of names: (husband, wife), where n is an integer less than or equal to 10. Have your program calculate and print out all seating arrangements where no husband is seated next to his wife.

18. In playing cards, you use a deck of 52 different cards. How many different hands of seven cards can be dealt?

19. The U.S. telephone system uses a telephone number consisting of ten digits. The first digit cannot be a 1; the second digit must be a 0 or a 1; and the remaining digits can be any digit at all. Calculate the maximum number of different phone numbers possible with this scheme.

20. As president of a new corporation, you are designing the company's tricolor flag.

a. If you are choosing from n colors, how many flags are possible using a color only once?

b. How many flags are possible if you can repeat a color but cannot put the same color on adjacent stripes?

c. How many flags are possible if the two outside colors are the same?

21. Show that the closed form solution for the number of derangements of n distinct elements generates the same set of numbers as the recurrence relation presented in this chapter.

22. PROGRAMMING PROBLEM: Write a program to do the following:

a. Read in two integers, n and r, which are less than 50.

b. Calculate the number of r—permutations without repetition of n objects.

c. Calculate the number of r—permutations of n objects with unlimited repetition.

23. Find the closed form solution to the following recurrence relations:

a. $A\ (n)\ =\ 2\ A\ (n-1)\ +\ n$

b. $B\ (n)\ =\ B\ (n\ /2)\ +\ 2\ n\ \log\ n$

c. $C\ (n)\ =\ C\ (n-1)\ +\ 3\ n^2$

24. Finish the proof of Theorem 6.10.

25. Show that $C\ (n,\ r)\ =\ C\ (n-\ 1,\ r)\ +\ C\ (n-\ 1,\ r-\ 1)$.

CHAPTER 7

GRAPHS AND THEIR USE IN COMPUTING

In computer science, it is often convenient to draw a picture that illustrates a relationship among objects. For instance, Venn diagrams are a useful tool for exploring properties of sets. We saw in the last chapter that a Venn diagram gave us a good idea of how the principle of inclusion/exclusion works. Properties that may be obvious from a diagram may not be evident otherwise. Because pictorial representations can reveal much about objects, we turn now to a study of graphs, and we examine how graphs can be used in computing.

WHAT IS A GRAPH?

Recall that in Chapter 0, we described a problem in which the city school district needed to determine the best bus route. A map of the bus stops and the connecting roadways is shown in Figure 7.1. The problem involved deciding which was the shortest sequence of roads to use in order to stop at each of the stops. Each stop is represented as a dot on the map, and the connecting roads are represented as lines connecting the dots. This illustration is an example of a discrete structure known as a *graph*. This kind of graph is different from the graphs you may have seen in your algebra classes. A graph can represent objects in terms of points and edges connecting some or all of those points. Often, each point represents some

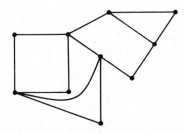

Figure 7.1

object, and each edge represents a relationship between points. In the example above, an edge is placed in the graph only when there is a direct route from one bus stop to another.

Formally, we define a **graph** G as an ordered pair of sets. The first element of the pair is a set P of **points** (or **vertices** or **nodes**). The second member of the pair is a set Q of unordered pairs of points of P. How does our map of the school bus stops fit into this definition? The bus stops make up the set P of points. Each road of the map connects bus stop a with bus stop b, so we can think of each line as an unordered pair { a, b }. In this way, the set Q defines all pairs related to one another by the relation defined as "has a direct route to." On our graph, then, each pair { x, y } of Q is called a **line** or **edge** of G. If we label the bus stops on our map as in Figure 7.2, then we can refer to a road by its endpoints. For instance, we can write

$$w = ab$$

to signify that line w is the line connecting point a to point b. We say that line w **joins** a to b and that a and b are **adjacent** points. We also say that line w and point a are **incident** with one another. If lines w and x are incident with a common point, the lines are said to be **adjacent lines**.

Thus, we can write this map as a graph G where the point set P is

$$\{ a, b, c, d, e, f, g, h, i, j \}$$

and the set of lines Q is

{ { a,b }, { a,i }, { b,c }, { b,g }, { c,d }, { c,e }, { g,i },
{ d,e }, { b,j }, { e,f }, { f,g }, { g,h }, { i,h }, { i,j } }

Note *hat there are 10 points in P and 14 lines in Q, so we say that $G = (P,$

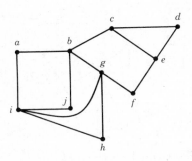

Figure 7.2

Q) is a $(10, 14)-$graph. In general, if $\mid P \mid$ is p and $\mid Q \mid$ is q, then the graph $G = (P, Q)$ is a $(p, q)-$ **graph**. The $(1, 0)-$graph is called the **trivial graph**.

Note that our definition of a graph allows us to have as an edge a pair of the form $\{ x, x \}$, for x an element of P. In other words, an edge can connect a point to itself. Such an edge would look like the one in Figure 7.3 and is known as a **loop**. When might an edge occur? Viewing the set of edges of a graph as a relation defined on the set of points, a loop at point x occurs whenever x is related to itself. Thus, a loop occurs at every point at which the relation is reflexive.

Some definitions of graph also allow multiple edges. These graphs are called **multigraphs**. Q can then have more than one pair of the same points, and the diagram corresponding to the graph has an edge for each occurrence of that pair. For example, suppose the graph G_1 were to consist of the sets

$$P = \{ a, b, c, d, e \}$$

and the edges

$$Q = \{ \{ a, b \}, \{ a, c \}, \{ b, c \}, \{ b, d \}, \{ b, d \}, \{ b, e \} \}$$

The graph's diagram looks like Figure 7.4, with two edges joining b and d. This type of graph might represent a bus stop map, and the duplicate edges between b and d would indicate that there are two different roads between stops b and d.

In our bus stop example, the points of the graph correspond to specific bus stops in the city. The labels a through j, then, correspond to the names of those bus stops. Such a graph is therefore called a **labeled graph,** and it is important to associate the labels with the points. In many other graphs, the labels are unimportant and are there only for easy reference when discussing properties of a graph. For example, the graph G_3 in Figure 7.5 has no labels. This type of graph is naturally called an **unlabeled graph**.

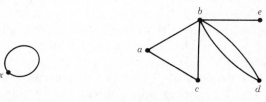

Figure 7.3   **Figure 7.4**

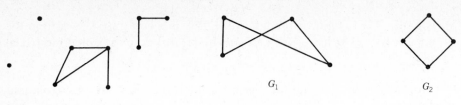

Figure 7.5   **Figure 7.6**

For a while, we will restrict our discussion to finite, unlabeled graphs with no loops or multiple edges. This means that all of our graphs $G = (P, Q)$ will have $| P |$ be finite. How can we tell when two of these graphs are the same? For instance, look at the graphs G_1 and G_2 in Figure 7.6. Try to visualize picking up the top leftmost point of G_1 and moving it down and to the right, aligned vertically with the top rightmost point, as shown in Figure 7.7. When you move the point, the edges attached to that point move with it. In some sense, graph G_1 is the same as graph G_2, even though at first glance they look different. These two graphs are said to be *isomorphic*. In general, two graphs G and H are **isomorphic** if there is a bijection from the set of points of G to the set of points of H which preserves the adjacency relationships. The isomorphism relation on graphs is an equivalence relation, and the proof of that is left for you as an exercise.

When two graphs G and H are isomorphic, we write $G = H$. There are certain properties that are common to a graph and all graphs isomorphic to it. For example, if G has p points and q lines, any graph isomorphic to G must also have p points and q lines. Such numbers associated with G which remain the same for any graph isomorphic to G are called **invariants**. One of the intriguing and as yet unsolved problems in graph theory is to determine a set of invariants that completely describes a graph and distinguishes it from all other graphs that are not isomorphic to it.

SUBGRAPHS

When we defined sets, we also looked at sets contained within sets, namely, subsets. In a similar fashion, we will consider subgraphs of a graph. Simply put, a **subgraph** of a graph G is a graph having all of its points and edges in G, with the restriction that for any edge in the subgraph, its endpoints must also be in the subgraph. For example, if G is the graph shown in Figure 7.8, then the graphs in Figure 7.9 are some of the subgraphs of G. A subgraph whose set of points is identical with the set of points of G is known as a **spanning subgraph**. Note that the spanning subgraph does not have to contain all of the edges of G as well. For instance, if G is the graph in Figure 7.10, then a spanning subgraph can be the graph in Figures 7.11 or 7.12. As you can see, a graph can have more than one spanning subgraph.

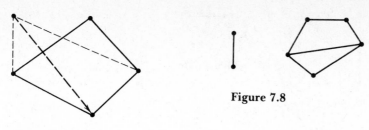

Figure 7.8

Figure 7.7

In our bus stop example, where each point is a bus stop and each edge a road from one stop to another, a spanning subgraph represents any map which includes all of the bus stops but not necessarily all of the edges.

A computer program can be represented as a graph. Let P be the set of statements in the program, and let Q be the set of edges defined in the following way:

> For points x and y in P, the edge { x, y } is in Q if and only if statement y can be reached directly from statement x.

Thus, computational statements in sequence would be represented as in Figure 7.13.

An IF—THEN—ELSE· construct would appear as Figure 7.14, and a GOTO, BRANCH, or JUMP might look like Figure 7.15. Using the graph of

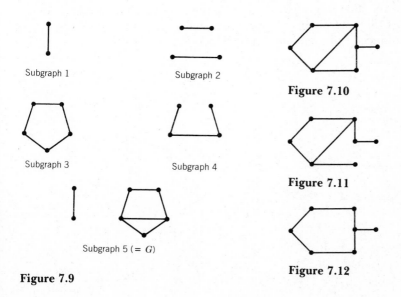

Subgraph 1

Subgraph 2

Figure 7.10

Subgraph 3

Subgraph 4

Figure 7.11

Subgraph 5 (= G)

Figure 7.9

Figure 7.12

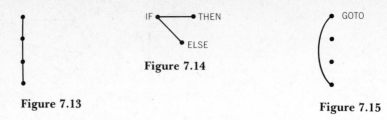

Figure 7.14

Figure 7.13

Figure 7.15

a program, we can see easily whether one statement in the program can be reached from another statement in the program. We can examine parts of the program (e.g., we can look at one procedure at a time) by considering only some of the points (that is, some of the statements) of the graph. Which edges should we include? We will include only those edges which connect any two points in the subset we have chosen.

More formally, suppose S is a subset of the set of points P of a graph G. Let us include in our new graph S the corresponding set of edges of G which join any two points of S. As an example, suppose G the graph in Figure 7.16 and S is the set $\{ a, b, d, g \}$. As you can see from the graph G, the edges

$$\{ \{ a, b \}, \{ a, d \}, \{ a, g \}, \{ d, g \} \}$$

are the edges of G which are incident with the points of S. Call this set of edges T. The subgraph of G defined by (S, T) is the graph induced on S by G and is denoted by $< S >$. In this case, then, $< S >$ looks like Figure 7.17. For any set S of points of G, the **induced subgraph** $< S >$ **of** G is the largest subgraph of G which contains exactly the set S as its set of points. It is easy to see that for any set S of points, there is a unique induced subgraph $< S >$ of G. In our example of the graph of a program, what does the induced subgraph of a procedure look like?

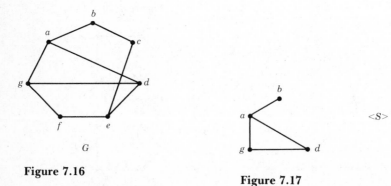

G

Figure 7.16

Figure 7.17

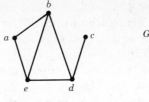

Figure 7.18

RECONSTRUCTING A GRAPH FROM ITS SUBGRAPHS

Suppose we have a graph G with point set P and edge set Q. For convenience, label the points of G as v_1, v_2, ..., v_n. For each i from 1 through n, we can form a subgraph of G in the following way. Let the set S_i be the set containing all of the elements of P except for point v_i. Then, let G_i be the induced subgraph $< S_i >$. This means that G_i has as its set of edges all of the edges of G which connect pairs of points of S_i. In a sense, then, G_i is the largest subgraph of G which does not contain v_i. For example, let G be the graph in Figure 7.18. G is a graph on five points, so we can form G_1, G_2, G_3, G_4, and G_5 as shown in Figure 7.19.

Can we work in the other direction? Suppose we don't know what G looks like, but we are presented with the graphs G_1 through G_5. Could we reconstruct G from the subgraphs? If so, the reconstruction process would have many useful "real world" applications. For example, a biologist may have information about pieces of a gene's chromosome structure. By representing that information as a set of subgraphs, we could reconstruct the original graph from them to view the complete gene. Unfortunately, even though the subgraphs contain a lot of information about the graph G itself, it is not yet known whether reconstruction is always possible. With graphs G_1 through G_5 above, we have the aid of knowing that the points are in the

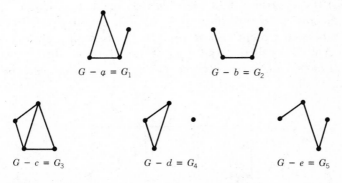

Figure 7.19

same relative position; we can simply lay one atop the other to reconstruct the original graph G. In general, however, if the points are not labeled, we don't know which point of one subgraph is the same as one point of another subgraph. Recall that in Figure 7.6, the two graphs are isomorphic but appear very different. Isomorphic copies can make reconstruction very difficult. In fact, it is not even known how to tell whether a set of subgraphs is really the complete set of induced subgraphs of a particular graph; that is, it is not known whether we can tell if the set of graphs presented to us is a set legitimately obtained by deleting successive points of a graph.

WALKS AROUND GRAPHS

In the study of graph theory and in the use of graphs in computer science, certain classes of graphs occur repeatedly. Let us look at some of them. Think of a graph as a map, so that the edges represent the ways in which we can travel from one point to another. In the graph G in Figure 7.20, for example, we can think of starting at point v_1 and traversing the edges and points in a variety of manners. For example, we can walk from v_1 along edge e_1 to point v_2, then along edge e_6 to point v_8, and so on. We can describe this walk as a sequence of points and edges. Formally, a **walk** in a graph G is an alternating sequence of points and edges

$$v_1, e_1, v_2, e_2, ..., v_{n-1}, e_n, v_n$$

This sequence must begin and end with a point, and each edge must be incident with the points before and after it in the sequence.

The points in a walk do not have to be distinct. In fact, if v_1 and v_n are identical, then the walk is called a **closed walk**. (Otherwise, the walk is an **open walk**.) If n is at least 3, if the walk from v_1 to v_n is closed, and if the points v_2 through v_n are distinct, then we call this a **cycle**. Informally, a cycle is a walk from a point back to itself that passes through no point more than once. A walk is called a **trail** if all of the edges are distinct; the walk is a **path** if all of the points are distinct. Let us look at some examples. In the

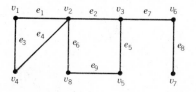

Figure 7.20

graph of Figure 7.20, the sequence

$$v_1, e_1, v_2, e_2, v_3, e_5, v_5, e_9, v_8, e_6, v_2, e_2, v_3$$

is a walk but not a trail because e_2 occurs twice. The sequence

$$v_2, e_1, v_1, e_3, v_4, e_4, v_2, e_6, v_8$$

is a trail but not a path. The sequences

$$v_1, e_1, v_2, e_2, v_3$$

and

$$v_1, e_3, v_4, e_4, v_2, e_2, v_3$$

are paths, and

$$v_1, e_1, v_2, e_4, v_4, e_3, v_1$$

and

$$v_2, e_2, v_3, e_5, v_5, e_9, v_8, e_6, v_2$$

are cycles. We will see shortly why these classes of graphs are important.

CONNECTEDNESS

So far, most of the graphs in our examples have been of graphs made up of one "piece." In other words, if we were to stand on a point in a graph, we could walk along the edges (i.e., along a "walk") to any other point in the graph. In some sense, then, the graphs have been connected. However, it is perfectly legal to have a graph made up of several "pieces," so that it is impossible to walk from a point in one piece to a point in another. More formally, we call a graph **connected** if, for every pair of points of the graph, there exists a path from one of the points to the other.

Let us look at the disconnected graph in Figure 7.21. Note that we have marked one point on one of the pieces as point x. Form the set Gx by following two steps:

1. First, let the point set Px be the set of all points of G reachable by a path from x.
2. Second, form the induced subgraph $< Px >$ of G.

Figure 7.21

What we end up with is the connected piece of G which contains x. This connected piece is called the **component of** G containing x. A subgraph is a component of a graph G if it is the largest connected subgraph containing itself.

A graph that is not connected must therefore have at least two components. The graph in Figure 7.21 has four components. When walking through a disconnected graph, you are limited by the component you are in. The **distance** between any two points in a graph is the length of the shortest path connecting the two points. If the two points are in different components of the graph, then the distance between them is said to be infinity.

Connectivity and distance are important when solving routing problems. For instance, it is crucial in our bus routing to know whether we can get from one point to another. Knowing that, we can turn to matters of distance between pairs of points.

DEGREE OF A POINT

On our bus routing map, we may have several roads leading out from our chosen starting place. In many graphs, there is more than one edge incident with a point. Thus, in a graph G, the number of edges incident with the point v is called the **degree** of the point v. There is a simple relationship which describes the edges of a graph and the degrees of its points.

THEOREM 7.1 _____

The sum of the degrees of the points of a finite graph with no loops or multiple edges is equal to twice the number of edges in the graph.

_____

To prove this, let G be a graph with p points and q edges. Each edge is incident with two points, so each edge contributes twice when we add up the degrees of the points. In other words, the edge is counted once for each endpoint, so we are double—counting the edges. Thus, we have

$$\sum_{v \text{ in } G} \text{degree}(v) = 2q$$

and hence we have proved our result. ∎

In particular, this theorem tells us the following:

COROLLARY **7.2**

In a finite graph with no loops or multiple edges, the number of points of odd degree is even.

To prove this, suppose there is an odd number of points of odd degree in the graph G. No matter how many points of even degree there are, the sum of the degrees of the points having even degree must be even. Further, with an odd number of points having odd degree, the sum of the degrees of the points having odd degree must be odd. Summing the degrees of all of the points must thus result in a sum which is odd. However, this contradicts Theorem 7.1, since the sum is a multiple of 2 and therefore even. ∎

REGULAR GRAPHS AND COMPLETE GRAPHS

Remember that we are considering only finite graphs with no loops or multiple edges. Let us turn our attention to graphs where the degree of any point is equal to the degree of any other point. A graph in which all the degrees are equal to a fixed number n is called a **regular graph of degree** n. The graphs in Figure 7.22 are regular graphs of degrees 1, 3, and 4. Regular graphs are not necessarily unique, since the two graphs of degree 4 are different. (They do not even have the same number of points.) Note, too, that a regular graph need not be connected.

Suppose we have a graph in which every point is connected by an edge to every other point. Such a graph is called a **complete graph** because the edges are completely filled in; there is no room to put in an extra edge. We denote the complete graph on n points as K_n. First, we note that the complete graph on n points is unique. (Why?) Second, it is easy to see that the complete graph on n points is connected and is a regular graph of

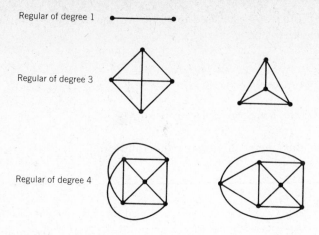

Regular of degree 1

Regular of degree 3

Regular of degree 4

Figure 7.22

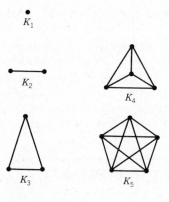

K_1

K_2

K_4

K_3

K_5

Figure 7.23

degree $n - 1$. The complete graphs on one point, two points, three points, four points, and five points are shown in Figure 7.23.

A computer network can be realized as a graph. We include in our graph a point for every computer in the network and an edge connects two points if the two computers can communicate directly with one another. If such a graph is complete, then every computer in the network can communicate with every other computer in the network. Exercise 11 will show you how many edges are in such a graph.

Suppose G is a graph that is not complete. What is needed to make G a complete graph? The **complement** G' of a graph G is the graph with the same set of points as G but with the edge set defined as follows:

Two points are adjacent in G' if and only if they are not adjacent in
G.

Thus, between a graph G and its complement G', we use up all possible
edges for the given point set. What is the complement of a complete graph
on n points? Since all of the edges are in the complete graph, its complement must consist of n points with *no* edges joining them. Given the graph
in Figure 7.24, its complement is in Figure 7.25.

Sometimes, a graph is its own complement. For example, the complement of the graph in Figure 7.26 is the graph in Figure 7.27. To show that
one graph is isomorphic to another, you must find a mapping from the
points of one graph to the points of other in which the edge relations are
preserved. Exercise 12 asks you to find a mapping to show that the graphs
in Figures 7.26 and 7.27 are isomorphic. In this case, that is, when a graph
is isomorphic to its own complement, the graph is called **self-complemen-
tary**.

RAMSEY NUMBERS

Let us look at a problem related to the complement of a graph. We can
state the problem as follows:

In any network of six computers, will there always be either three
computers all of which can communicate directly with one another,
or three computers none of which can communicate directly with
one another?

Is the answer to this question "yes" or "no"? Let us represent the situation
by drawing a graph. Each point will represent a computer in the network. If
two points are adjacent, the two computers can communicate with one
another. Then we can restate the problem as follows:

In any graph on six points, must it be the case that either there are
three points all of which are adjacent to one another or three points
none of which are adjacent?

Figure 7.24   Figure 7.25   Figure 7.26   Figure 7.27

In other words, in any graph on six points, is there always a subgraph which is a complete graph on three points or three points which are totally isolated from one another? Because the complement of a graph contains a line between two points whenever the original graph does not, this means that we want to prove the following:

THEOREM 7.3 ───

In any graph on six points, either the graph or its complement contains a complete graph on three points.

───

To prove this, let's note that the complete graph on three points is just a triangle, as shown in Figure 7.28. Let v be one of the six points of a graph G. We can assume that v is adjacent to at least three points in G. (If not, v must be adjacent to at least three points in G', and we can reverse the roles of G and G' in this proof.) Let those three points be v_1, v_2, and v_3. If any two of the v_i are adjacent to each other (say, v_1 and v_2), then $v_1 - v_2 - v$ makes up the required triangle. If no two of them are adjacent, then $v_1 - v_2 - v_3$ must be a triangle in G'. ■

───

This theorem is a special case of a more important result known as Ramsey's Theorem. **Ramsey's Theorem** states that for each pair of positive integers m and n there is a smallest integer $R(m, n)$ such that every graph on $R(m, n)$ points must contain a complete graph on m points or its complement must contain a complete graph on n points. Finding a general algorithm to compute the Ramsey number $R(m, n)$ for any possible m and n is an unsolved problem. (Of course, some values of $R(m, n)$ are known.) The existence of the Ramsey numbers is useful in many proofs where we are looking for a minimal set of things which satisfy a certain condition.

BIPARTITE GRAPHS

Let us look at a few more special kinds of graphs. In computer science, we often have problems in which we must do some sort of matching. We may

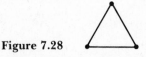

Figure 7.28

have two different sets, and we want to pair the elements of the first set with the elements of the second in some fashion. A simple example of this is the performance of a port controller: it must pair incoming ports with outgoing ones. A graph which represents this kind of matching is called a *bipartite graph*. In a **bipartite graph,** the set P of points can be partitioned into two distinct set P_1 and P_2, so that every edge of the graph joins a point of P_1 with a point of P_2 . If a bipartite graph contains every edge possible from P_1 to P_2, then the graph is a **complete bipartite graph** on m and n points, where $| P_1 | = m$ and $| P_2 | = n$. As a shorthand, we write this graph as $K_{m,n}$. It is easy to see that the complete bipartite graph on m and n points has mn edges.

Look at the bipartite graphs in Figure 7.29. Are there any triangles in any of them? No. Let's see why.

THEOREM 7.4 _____

A graph is a bipartite graph if and only if all of its cycles are even.

_____

To prove this, let G be a bipartite graph with points partitioned into sets P_1 and P_2, so that every edge of G joins a point of P_1 with a point of P_2. Let C be a cycle in G, and let x be a point on the cycle. Since the labels P_1 and P_2 are chosen arbitrarily, call P_1 the set containing x. Then, the cycle must go from x in P_1 to a point x_1 in P_2, then to x_2 in P_1, to x_3 in P_2, and so on, until returning to x in P_1. Thus, the points x_i with i even must be in P_1 and those with i odd must be in P_2. It is clear, then, that the length of the cycle must be even.

For the converse, suppose G is a graph all of whose cycles are even. Let us assume that G is connected. (If not, then this argument can be repeated for each component of G.) Let x be any point in G, and let P_1 be the set consisting of x and of all points of G whose distance from x is even. (Remember that the distance between two points is the length of the shortest path connecting them.) Let P_2 be the set of points of G which are not points in P_1. Suppose there is

$K_{2,3}$

$K_{1,2}$

$K_{3,3}$

Figure 7.29

an edge of G which joins a point of P_1 with another point of P_1. Let y and z be the endpoints of this edge. Then, the shortest path from x to y, united with the shortest path from x to z and the edge from y to z, form an odd cycle. This contradicts the fact that all the cycles of G are even. A similar argument holds had we tried to find an edge joining two points in P_2 instead of P_1. Because neither component contains an edge between two points in that component, every edge of G must connect a point in P_1 with a point in P_2. ∎

The bipartite graph appears in problems concerning the handling of sparse matrices on the computer. Most of the entries in a sparse matrix are zero. There are techniques for storing sparse matrices that reduce the amount of storage space required, and these techniques relate to reducing a graph from the size of a complete graph to one which is a bipartite graph.

TREES

We say that a graph is **acyclic** if it has no cycles, and a connected, acyclic graph is called a **tree**. A graph without cycles is a **forest** ; thus, a forest is a graph whose components are trees. The concept of a tree pervades computer science, and it is especially useful for handling data in a computer. Many sorting and searching techniques require the traversal of a tree. Thus, let us examine in some detail the properties of trees.

There are several equivalent ways to define a tree.

THEOREM 7.5 _____

Let G be a graph with p points and q edges. The following statements are equivalent:

1. G is a tree.
2. G is connected and $q = p - 1$.
3. G is acyclic and $q = p - 1$.
4. G is acyclic, and the addition of another edge forms a cycle.
5. G is connected but the deletion of any edge disconnects G.
6. For any two points v_1 and v_2 in G, there is exactly one path from v_1 to v_2.

First, we will show that statement 1 implies statement 2. Let G be a tree with p points and q edges. Then, G is connected and acyclic.

We will use induction to show that statement 2 is a result of statement 1. As a basic step, if p is equal to 1, then G must be a single point, so $q = 0 = p - 1$. If p is equal to 2, then the connectivity of G implies that G is a single edge. Thus, again $q = p - 1$. For the inductive step, let us assume that any tree with n points is connected and has $n - 1$ edges, for n less than or equal to k. Let G be a tree with $k + 1$ points. If we remove an edge of G, we must end up with two graphs G_1 and G_2 that are trees, since they are still connected and acyclic. Further, if G_1 has p_1 points and G_2 has p_2 points, we know that

$$p_1 + p_2 = k + 1$$
$$p_1 < k + 1$$

and

$$p_2 < k + 1$$

Thus, G_1 must have $p_1 - 1$ edges, and G_1 must have $p_2 - 1$ edges. Hence, adding back in the edge we removed from G originally, we have

$$q = 1 + (p_1 - 1) + (p_2 - 1)$$
$$= (p_1 + p_2) - 1$$
$$= (k + 1) - 1$$

edges, and our result is proved by induction.

We have thus shown that the first definition implies the second. To show the equivalence of all of these definitions, we will use a proof technique that is different from what you have seen previously in this book. We will show that definition 2 implies definition 3, 3 implies 4, 4 implies 5, 5 implies 6, and 6 implies definition 1. Thus, we will show that the statements are like a cycle in a graph: there is a path from any one to any other one and another path back to the first one. Therefore, all six are equivalent.

Definition 2 implies definition 3:

We want to show that if a graph G is connected and $q = p - 1$, then G is acyclic and $q = p - 1$. Suppose G is connected and $q = p - 1$. Assume that the opposite of statement 3 is true; that is, assume G has a cycle on n points, for some n less than p. Then, there are n points on the cycle and thus n edges. For each of these $p - n$ points not on the cycle, there is an edge on the shortest path from that point to a point on the cycle. Each of these edges must be different, which means that

$$q \geq n + (p - n) = p_1$$

so q is at least as large as p. This is a contradiction, so G must be acyclic.

Definition 3 implies definition 4:

Now, suppose G is acyclic and $q = p - 1$. We want to show that G is connected, it is acyclic, and the addition of another edge forms a cycle. Since G is acyclic, each component of G must be a tree. Let k be the number of components of G. Then, each component has one more point than edge, so we must have

$$p = q + k$$

However, we know that $p = q + 1$, so k must be 1; therefore, G is connected. Let x and y be any two points which are not adjacent in G. We want to add the edge joining points x and y [denoted by (x,y)] to G. However, since G is connected, there must be a path connecting x to y. By adding the edge (x, y), we now have two different paths connecting x and y, so there is a cycle containing x and y, and our assertion is proved.

Definition 4 implies definition 5:

Now, assume that G is acyclic and that the addition of another edge forms a cycle. We want to show that G is connected but the deletion of any edge disconnects G. First, suppose G is not connected. Then, G must have at least two components. Let x be a point in component C_1 of G and y be a point in component C_2 of G. Then, there is no path from x to y, since they are in different components. Let us add the edge (x, y) to G. Our assumption tells us that this forms a cycle containing x and y. However, the cycle must be made up of the edge (x, y) and a path connecting x to y. This cannot be, unless x and y are in the same component, which contradicts our assumption. Thus, G must be connected.

Now, let e be an edge of G with endpoints v_1 and v_2. Suppose we remove e from G to form a new graph H. If H were connected, there would be a path in H from v_1 to v_2. However, this path is also in G, and the addition of e to the path forms a cycle in G. But we know that G is acyclic! Thus, the removal of an edge from G must disconnect G.

Definition 5 implies definition 6

Assume that G is connected and that the deletion of any edge disconnects G. We want to show that there is exactly one path between any two points of G. The fact that G is connected means that there is at least one path between any two points. Suppose v_1 and v_2 are points of G connected by two distinct paths. Let P_1 be the first path and P_2 be the second, as shown in Figure 7.30. Remove an edge from P_1. Call that edge e, and let the endpoints of e be x and y. By our assumption, this should disconnect G. However, we can travel from x along P_1 to v_1, then from v_1 to v_2 along P_2, and finally back to y along P_1, so x and y (and hence G) must still be connected. This contradicts our assumption, so it must be the case that there can be at most one path between any two points in G.

Definition 6 implies definition 1:

Suppose there is exactly one path from any point in G to any other point in G. We want to show that G is connected and acyclic. The fact that at least one path exists between any two points of G tells us that G is connected. Suppose G contains a cycle. Let x and y be points on the cycle. Then, the cycle defines two distinct paths between x and y, which contradicts the fact that there is at most one such path. Hence, G must be acyclic. This completes our proof. ■

For any finite connected graph G, we can find a tree whose point set is the same as the point set of G and whose edges are a subset (not necessarily a proper subset) of the edges of G. For instance, if G is the graph in Figure 7.31, then the tree in Figure 7.32 is a subgraph of G that has the same point set as G. Such a tree is called a **spanning subtree of** G. Can there be more than one spanning subtree of a graph? The answer is left for you to ponder as an exercise. We will return to spanning subtrees when we learn about networks and network flows.

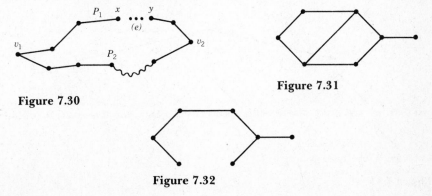

Figure 7.30

Figure 7.31

Figure 7.32

ROOTED TREES

There is a particular kind of tree that is used very often in computer science. We mentioned before that trees are involved in many sorting and searching algorithms. Let us look at an example. Suppose you are searching through the directory of your computer system. Your computer is likely to have a naming convention similar to the one depicted in Figure 7.33. This tree really describes the names of the files in the directory. To find a particular file, you follow the branches of the tree. You always begin at the top of the tree and work your way down. Thus, the point of the tree which is at the top is distinguished from the rest of the tree. We call this point the **root** of the tree. In general, a tree is a **rooted tree** if one point of the tree is distinguished from all of the others. Not all trees are rooted, although one can always designate a point as the root. Spanning trees, for example, are usually unrooted trees. However, most applications of rooted trees use directed graphs (digraphs) rather than undirected graphs. As we shall see later, it is much easier to distinguish a point from all others in a digraph (e.g., the root can be the point that has no edges leading into it). In an undirected graph, there is often no obvious method of defining some point which will stand out as being distinguishable from another point.

Rooted trees are used in sorting and searching and in many other algorithms. One application with which you may be familiar is in defining programming languages.

The *Backus Naur form* (also called Backus Normal form, or BNF) is a convention which uses the following kinds of notation:

> Individual symbols
> A separator symbol (|)
> A definition symbol (:=)
> Class names (<name>)

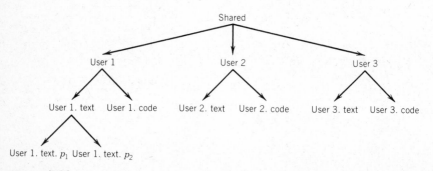

Figure 7.33

For instance, in the ALGOL language, defining parts of a number involves defining the following classes:

<digit>:= 0 | 1 | 2 | 3 | 4 | 5 | 6 | 7 | 8 | 9
<unsigned integer>:= <digit> | <unsigned integer> | <digit>
<integer>:= <unsigned integer> | + <unsigned integer> |
− <unsigned integer>
<decimal fraction>:=. <unsigned integer>
<decimal number>:= <unsigned integer> | <decimal fraction> |
<unsigned integer> <decimal fraction>

We can draw a precedence graph to reflect the definitions:

<digit>

<unsigned integer>

<integer>

<decimal fraction>

<decimal number>

In this case, all definitions must begin from the top, so the class <digit> must be the root of this tree. When using the Backus Naur form, symbols and classes can be concatenated. For example, we may have

<integer>:= <integer> + <integer>

Note, too, that many of the definitions have the same symbol appearing on both sides of the:= sign. Thus, they are recursive definitions.

You will see in the next chapter that derivation trees for grammars are also rooted trees.

PLANAR GRAPHS

Now that we have defined several types of graphs, let us turn to an examination of some of the properties of graphs. Recall that in the section on logic design, we were concerned about the number of crossovers when designing circuits. It is easy to see how the circuit diagrams can be viewed as graphs, and we want to make sure that there are as few crossovers as

Figure 7.34   Figure 7.35

possible. For some graphs, it is always possible to draw the edges and points so that two edges intersect only at endpoints of the edges. However, for many graphs, it is impossible to make such a drawing. For instance, try to draw the complete graph on six points without having the edges cross over one another. We call a **crossover** a place in the drawing of the graph where the edges intersect at a place other than the endpoints. The crossovers can be important in considering printed circuit boards because an extra layer of insulation is needed between edges that cross over one another. Not only does the insulation add to the cost of the printed circuit board, but it also adds to the thickness of the board.

We say that a graph is **planar** if it can be drawn on the plane without having any two edges intersect. The phrase "it can be drawn" means that a particular drawing of a graph may have crossovers, but at least one isomorphic copy of the graph has no crossovers. If a graph is planar, it is isomorphic to a graph that has no crossovers. For example, the graph in Figure 7.34 has a crossover, but it is isomorphic to the graph in Figure 7.35 which has none. To distinguish one from the other, we will call a graph embedded in a plane (i.e., one drawn without crossovers) a **plane graph**.

A plane graph determines various regions of the plane. For instance, the graph in Figure 7.35 defines four "regions" or "spaces" of the plane, including the unbounded region on the "outside" of the graph. The number of regions is related to the points and edges of the plane graph by the following formula:

THEOREM 7.6 _____

For a plane graph with p points, q edges, and r regions,

$$p - q + r = 2, \text{ or } r = q - p + 2$$

_____

The inductive proof is left as an exercise. ∎

_____

This theorem is a variation of Euler's polyhedron formula which you may have used in your high school geometry class. The theorem can be quite useful. One of its applications is in finding an easy way to tell whether a graph is planar or not. To do this, first we prove that two particular graphs are not planar.

LEMMA 7.7 _____

If G is a planar graph with p points and q edges and $p \geq 3$, then q must be less than or equal to $3p - 6$.

_____

Since G is a planar graph, assume that you are looking at its equivalent plane graph. Think about adding edges to G without having the new graph lose its planarity. The smallest number of edges in a plane graph needed to define a region is 3, so the maximum number of edges in a plane graph would occur when each region is a triangle. In this case, each edge of the new graph is incident with two regions, and each region has three edges. This means that $3r = 2q$. Substituting this in Theorem 7.7, we see that the maximum number of lines must be $3p - 6$. ■

_____

This lemma gives us the ammunition to prove the following theorem about the graphs in Figure 7.36.

THEOREM 7.8 _____

The graphs K_5 and $K_{3,3}$ are not planar.

_____

The complete graph on five points has 10 edges, so q is larger than $3p - 6$. For $K_{3,3}$, note that every region is made up of a cycle of 4 edges. Using an argument similar to the one for Lemma 7.7, we can show that a planar graph in which every region is a four—cycle can have at most $2p - 4$ edges. (This is left for you as an exercise.) In $K_{3,3}$, we have six points and nine edges, but $2p - 4$ is equal to 8. ■

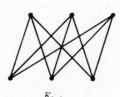

K_5 $K_{3,3}$

Figure 7.36

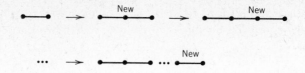

Figure 7.37

Finally, let us introduce one more concept that will enable us to characterize planar graphs. We say that two graphs are **homeomorphic** if both are
derived from the same graph by a sequence in which edges are subdivided.
For instance, any path is homeomorphic to any other path; we can take the
smaller path and keep adding points to an edge until we have the same
number of edges as the larger path, as shown in Figure 7.37. Likewise, any
two cycles are homeomorphic, as shown in Figure 7.38. This concept allows
us to state **Kuratowski's Theorem,** which says that a graph is planar if and
only if it has no subgraph homeomorphic to K_5 or $K_{3,3}$.

The proof is beyond the scope of this text. However, the result is
powerful, because it makes it easy for us to see whether a graph is planar or
not. For example, the "Petersen" graph, shown in Figure 7.39, is not planar
because it is homeomorphic to K_5.

TRAVERSABILITY: EULERIAN GRAPHS

Chapter 0 described a marathon in which the race administrators wanted
the running route to run through the town, crossing each of the city's
bridges exactly once. This problem is derived from a real—life problem
which confronted Leonhard Euler in 1736. The townspeople of the city of

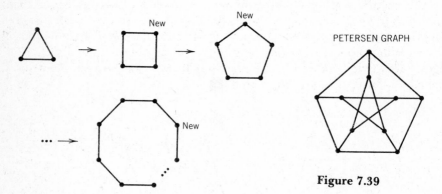

Figure 7.38

Figure 7.39

Königsberg, then in East Prussia, liked to go out for a Sunday walk. The river Preugel flows through Königsberg, and Euler wondered whether there was a route through the town that crossed each of the town's seven bridges exactly once. In solving this problem, Euler demonstrated a very easy way of deciding whether a graph has a path that traverses each edge exactly once.

First, let us redraw the picture of the city as shown in Figure 7.40 as a graph. We let a point represent each land mass, and we draw an edge from one point to another if there is a bridge connecting those two land masses. The result is the graph of Figure 7.41. We define an **Eulerian trail** as a trail which traverses each edge of a graph exactly once while going through all of the points. Thus, Euler was wondering whether this graph has an Eulerian trail. A graph is called an **Eulerian graph** if and only if it has an Eulerian trail. Euler answered the question by proving the following theorem.

THEOREM **7.9** _____

A connected graph G is Eulerian if and only if every point of G has even degree.

_____

Suppose G is a connected graph which is Eulerian. Then, there is a trail T which is an Eulerian trail. Whenever a point of G is on the trail T, there are two edges of T incident with that point, and there will be two edges for each time the point is on the trail. However, all of the edges of G are on the trail T, so each point of T must have even degree. The proof of the converse is left as an exercise. ∎

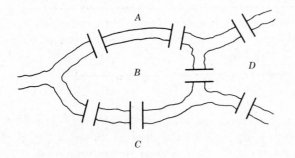

Figure 7.40

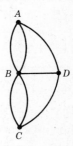

Figure 7.41

Is the map of Königsberg Eulerian? Theorem 7.9 tells us no, because there is at least one point of odd degree. Thus, any walk through the town must traverse at least one bridge twice. What is the minimum number of duplicate bridge crossings necessary?

HAMILTONIAN GRAPHS

A similar concept is that of finding a cycle that passes through every point of a graph. For instance, we can solve our bus route problem if we find a cycle that passes through every bus stop. In a previous section, we said that a subgraph of a graph is a spanning subgraph if every point of the subgraph is also a point of the original graph. In a similar manner, we say that a cycle of a graph is a **spanning cycle** if the cycle is a spanning subgraph. A graph is **Hamiltonian** if it has a spanning cycle. The concept is named after Sir William Hamilton, who first investigated the consequences of this property. If such a spanning cycle exists, it is called a **Hamiltonian cycle**. It was fairly easy to characterize graphs as Eulerian (see Theorem 7.9), but there is no easy characterization of Hamiltonian ones. The best we can do is state a result of Posa's:

THEOREM 7.10 _____

Let G be a graph having $p \geq 3$. If for every n such that

$$1 \leq n < (p-1)/2$$

the number of points of G of degree exceeding n is less than n, and if, for p an odd number, the number of points of degree $(p-1)/2$ does not exceed $(p-1)/2$, then G is Hamiltonian.

_____

We will not present a proof of this theorem here. The statement of this theorem is complex, but there are several weaker corollaries that act as easy tests of a graph's being Hamiltonian.

COROLLARY 7.11

If G is a graph on at least three points, and if for every pair of nonadjacent points x and y, the sum of the degrees of x and y is at least the number of points of G, then G must be Hamiltonian.

_____

A similar result can be stated as follows:

COROLLARY 7.12

If G is a graph on at least three points and if the degree of every point of G is at least $p/2$, then G must be Hamiltonian.

For example, the graph in Figure 7.42 has six points, and each point is of degree 3, so it must be Hamiltonian.

Let us consider an example of where knowing a Hamiltonian cycle may be helpful. Suppose you have an intricate computer network set up, with nodes of the network spread all over the world. The network can be represented as a graph: the points of the graph are the nodes of the network; an edge joins two points if the corresponding nodes can communicate directly. Once a month, a status report is written at each node of the network, and the reports are appended to one another to form one large network—wide status report. If we could find a Hamiltonian cycle through this network, each node could forward its part of the status report to the next node in the cycle, and by the time the end of the cycle were reached, the entire status report would be complete and in the hands of the person who submitted the initial section of the report.

COLORING A GRAPH

It is no coincidence that many of the problems we have discussed have related graphs to maps. In the Königsberg bridge problem and the school bus problem, we transformed a map into a graph in order to answer a question about the map. In a similar fashion, another question about maps arises which can be approached from a graph theory perspective. You are familiar with maps of the country and maps of the world. In most instances, these maps are colored with several colors so that it is easier for the viewer to distinguish one political region from another. The larger the number of colors in the map, the more expensive it is to print the map.

Figure 7.42

Thus, the obvious question is: what is the smallest number of colors needed to color a map so that adjacent countries are colored with different colors? Let us see how to transform this into a graph theory problem and then solve it. Look at the map in Figure 7.43. We can draw this map as a graph by letting a point represent each country, and drawing an edge from one point to another if and only if the corresponding countries are adjacent. Here, we consider two countries to be adjacent if and only if they share a common border. Our map becomes the graph in Figure 7.44.

Now, our problem can be restated in terms of the graph: what is the smallest number of colors needed to color the points of a graph so that adjacent points are colored with different colors? To answer this question, let us define some terms. A **coloring** of a graph is an assignment of colors to the points of a graph so that no two adjacent points have the same color. An $n-$**coloring** of a graph is a coloring using n colors. Thus, our question searches for the smallest n which will always work. Note that an $n-$coloring partitions the point set P of the graph into n color classes. We say that the **chromatic number** $\chi(G)$ of a graph G is the minimum n for which G has an $n-$coloring. It is clear that any graph G has a $p-$coloring, where p is the number of points of G. Therefore, any graph G must have an $n-$coloring for each n between $\chi(G)$ and p. Furthermore, the complete graph on n points requires n colors.

One easy result that follows from these definitions is shown below.

THEOREM **7.13**_____

A graph G is bicolorable (i.e., it has a 2$-$coloring) if and only if G is bipartite.

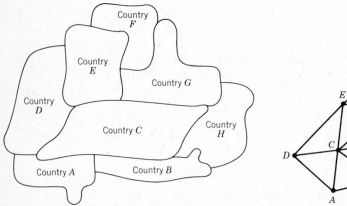

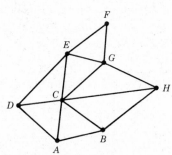

Figure 7.43   **Figure 7.44**

Recall that Theorem 7.4 stated that a graph G is bipartite if and only if all cycles are even. If G is bicolorable, then examine a coloring of G with two colors, say red and blue. Let P_1 be all the points in G which are colored red, and let P_2 be all of the points of G colored blue. Because no red point can be adjacent to another red point (and likewise blues), G must be a bipartite graph $K\,p_1,\,p_2$, where $\mid P_1 \mid\, = p_1$ and $\mid P_2 \mid\, = p_2$. Hence, all cycles are even. To prove the converse, suppose G has no odd cycles. Then G must be bipartite, so we can partition the points of G into two classes such that each point of the first class is adjacent only to points of the second class, and vice versa. Color all points of the first class with one color, and then color all of the points of the second class with a second (different) color. This is a bicoloring of G. ∎

Our mapmaker's problem can be refined somewhat. Every map of the world or of a particular country corresponds to a graph that can be drawn in a plane. (Why?) Thus, we want to find the smallest number of colors that will color any planar graph. It is easy to show that three colors are not enough, because the graph in Figure 7.45 requires at least four colors. It is also not too difficult to show that five colors is enough.

THEOREM 7.14_____

Every planar graph can be colored with five colors.

We will prove this theorem by using induction on p, the number of points of a graph. If a graph has five or fewer points, then the result follows immediately (because we showed above that every graph is p−colorable). Thus, as the inductive hypothesis, assume that any planar graph with p points (for p at least 5) is five−colorable. Suppose G is a planar graph with $p+1$ points. G must have a point of degree 5 or less. (If not, all points would be of degree 6 or more, and K_5 would therefore be a subgraph of G. However, we showed in Theorem 7.8 that this would force G to be nonplanar.) Let x be a point of degree 5 or less. Form a new graph H by

Figure 7.45

removing x and all the edges incident with x from G. This new graph H has p points, so it must be colorable with five colors., Suppose we call these five colors c_1 c_2, c_3, c_4,and c_5, and we examine a five—coloring of H. If there is some color c_i that is not assigned to one of the (at most five) points that were adjacent to x in the original graph G, then we can use c_i to color x in order to produce a five—coloring of G. Thus, the only case left to consider is the one where x is of degree 5, and these five points are colored with five different colors. Then, the configuration around x must look like Figure 7.46. Assume that point x_i is colored with color c_i. Let $H_{1,3}$ be the subgraph of H induced by those points of H which are colored with colors c_1 or c_3. $H_{1,3}$ may not be connected.

Case 1: If x_1 and x_3 belong to different components of $H_{1,3}$, then we can create a five—coloring of H by exchanging the colors of the points in the component of $H_{1,3}$ containing x_1. In this coloring, no point adjacent to x is colored with color c_1, so by coloring x with c_1, we obtain a five—coloring of G.

Case 2: Suppose, though, that x_1 and x_3 belong to the same component of $H_{1,3}$. Because $H_{1,3}$ is connected, there must be a path in G between x_1 and x_3 where all the points of the path are colored with c_1 or c_3. Combine this path with the path x_1-x-x_3. What results is a cycle which encloses either the point x or both of the points x_4 and x_5. In either of these situations, there can be no path which joins x and x_4 all of whose points are colored with colors c_2 or c_4. Let $H_{2,4}$ be the subgraph of H induced by the points colored with c_2 or c_4. Then, x and x_4 must belong to different components of $H_{2,4}$. We can exchange the colors of the points of $H_{2,4}$ containing x. What results is a five—coloring of H in which no point adjacent with x is colored with color c_2. We can then create a five—coloring of our original graph G by coloring x with color c_2. ∎

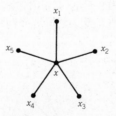

Figure 7.46

Since three colors are not enough and five will suffice, we have narrowed down colorability to two numbers: a planar graph is colorable either in four or five colors. Until recently, it was not known whether five colors was a necessity; that is, it was not clear that four colors would suffice to color all graphs. Two mathematicians at the University of Illinois answered this question by establishing that all possible maps can be reduced to approximately 1300 special cases. Then, they examined all of the cases using a computer program and thus proved the **Four Color Theorem**, which says that every planar graph is four—colorable.

REPRESENTING GRAPHS AS MATRICES

The proof of the Four Color Theorem raises an obvious question: how can a graph be represented on a computer? A graph can be represented as a matrix in several ways. To represent any graph, first we must label the points and edges of the graph. Suppose, then, that the points of a graph G are labeled x_1, x_2, ..., x_p, and the edges are labeled e_1, e_2, ..., e_q. The **adjacency matrix** of the graph G is formed by letting entry a_{ij} equal 1 if point x_i is adjacent to point x_j. If x_i and x_j are not adjacent, then a_{ij} is 0. For example, the graph in Figure 7.47 has the adjacency matrix

$$
\begin{array}{cccccc}
0 & 1 & 0 & 0 & 0 & 0 \\
1 & 0 & 1 & 0 & 1 & 0 \\
0 & 1 & 0 & 1 & 1 & 0 \\
0 & 0 & 1 & 0 & 0 & 0 \\
0 & 1 & 1 & 0 & 0 & 1 \\
0 & 0 & 0 & 0 & 1 & 0 \\
\end{array}
$$

In other words, if point x_i is adjacent to point x_j, then we put a 1 in the entry in the ith row and the jth column. Likewise, since x_j is adjacent to x_i, there is also a 1 in the entry in the jth row and the ith column. What does it mean if there is a 1 in an entry of the form x_{ii}? Such an entry would represent a loop. How would a multiple edge be represented in an adjacency matrix? That depends on the purpose to which the adjacency matrix

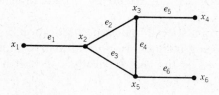

Figure 7.47

would be put. In some cases, the entry a_{ij} of the adjacency matrix is the number of edges connecting points x_i and x_j. In other cases, you may want a_{ij} to be restricted to 1s and 0s, no matter how many multiple edges you may have.

In this fashion, the description of the graph can be stored in a computer. Note that the adjacency matrix for a graph is always symmetric.

There are other matrices that can be defined. An **incidence matrix** of a graph G is one in which the entry a_{ij} is set to 1 if point x_i is incident with edge e_j; it is set to 0 otherwise. For the graph in Figure 7.47, the incidence matrix looks like this:

```
1 0 0 0 0 0
1 1 1 0 0 0
0 1 0 1 1 0
0 0 0 0 1 0
0 0 1 1 0 1
0 0 0 0 0 1
```

There are many other ways of describing graphs using matrices. For instance, a **path matrix** can be defined so that a_{ij} is equal to 1 whenever there is a path connecting point x_i to point x_j. Or, we can define a **cycle matrix,** wherein there is a row for each cycle and a column for each edge of the graph. The entry a_{ij} is 1 if the ith cycle contains edge e_j of the graph; it is 0 otherwise. The type of matrix used depends on the kind of problem that is being addressed and thus the properties of the graph involved in the problem.

THE PATH MATRIX AS TRANSITIVE CLOSURE

Let us examine in detail some of the properties of an adjacency matrix. Let D be the adjacency matrix for a graph G. D has a 1 in entry d_{ij} whenever there is an edge in G connecting point x_i to point x_j . An edge is a path of length 1, so we can also think of D as a type of path matrix which illustrates those points of G which are connected by a path of length 1.

Suppose we perform a special matrix multiplication on D by multiplying D by itself in the following manner. In normal matrix multiplication, we say that matrix $C = [\ c_{ik}\ ]$ is the matrix product $A \times B$ if

$$c_{ik} = \Sigma\ a_{ij} \times b_{jk}$$
$$= a_{i1} \times b_{1k} + a_{i2} \times b_{2k} + \dots + a_{in} \times b_{nk}$$

where $A = [\, a_{ij}\,]$, $B = [\, b_{jk}\,]$., $i = 1, ..., m$, $j = 1, ..., n$, and $k = 1, ..., s$. Thus, A is an $m \times n$ matrix, and B is an $n \times s$ matrix. The summation is from $j = 1$ to n. We want to multiply the adjacency matrix D by itself to form D^2. However, we want to use Boolean operators instead of the usual multiplication and addition. Thus, we form D^2 according to the formula above, but we use the Boolean AND instead of $\times$ and the Boolean OR instead of $+$. The resulting matrix D^2 has an entry of 1 in position d_{ij} when there is a path exactly of length 2 from point x_i to point x_j. Further, if we form the matrix D^2 OR D by using the Boolean OR to combine the ij—th entry of D^2 with the ij—th entry of D, our new matrix has a 1 in the ij—th position whenever there is a path of length 1 or 2 from point x_i to point x_j.

We can continue this process, forming first D^3, then D^3 OR D^2 OR D, then D^4 and D^4 OR D^3 OR D^2 OR D, and so on. In general, matrix

$$D^r \text{ OR } D^{r-1} \text{ OR... OR } D^2 \text{ OR } D$$

has a 1 in the ij—th position whenever there is a path of length at least 1 and no greater than r between point x_i and point x_j. Because G is a finite graph, we know that this process must terminate. (See Exercise 20.) Let D^* be the matrix in which this process terminates. What does D^* represent? If there is a 1 in position ij, then there is a path in G which connects point x_i with point x_j.

Suppose we define a relation R on the set of points of G so that $x_i \, R \, x_j$ whenever the points are adjacent; then the matrix D represents the relation R. Further, the matrix D^* represents the transitive closure of R. (Why?) However, D^* is also the *path matrix* of the graph G! It is for this reason that we say that the path matrix represents the transitive closure of the adjacency matrix of a graph.

DIRECTED GRAPHS

In all of the sections above, we have dealt with graphs whose edges were unordered pairs of points. However, there are many instances in which it is convenient or necessary to assign a direction to an edge. This corresponds to assigning an order to the pair of points which defines the edge. A **directed graph** (or **digraph** or **network**) is an ordered pair (P, Q), where the set P is a finite set of points and Q a collection of ordered pairs of points. The edges of a directed graph are sometimes called **arcs** or **directed lines**.

We can think of the edges of undirected graphs as two—way streets; an arc in a directed graph is thus a one—way street. Multiple arcs are allowed

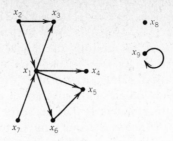

Figure 7.48

in directed graphs, and we must know the direction of travel from one endpoint to the other. Given two points a and b, the arc which begins at a and ends at b is different from the arc from b to a. When we considered undirected graphs, the degree of a point was the number of edges incident with that point. With directed graphs, we break the degree down into two components: in−degree and out−degree. The **out−degree of a point** x is the number of arcs incident with x for which x is the initial point of the arc. In other words, the in−degree of x counts the number of arcs originating from x or leaving x. The **in−degree of a point** x is the number of arcs incident with x for which x is the final point of the arc. Thus, the in−degree of x counts the number of arcs coming in to x from other points. In Figure 7.48, the out−degree of x_1 is 4, the out−degree of x_1 is 2, the in−degree and out−degree of x_8 are both 0, and the in−degree and out−degree of x_9 are each 1.

The definitions of many of the other concepts of undirected graphs carry over to directed graphs with little change. For example, walks, cycles, and paths are the same with obvious restrictions. A path *from a to b* implies directed arcs *from* point a *to* point b and *from* point b *to* point c. One thing that changes is the definition of an adjacency matrix. In an adjacency matirix for a digraph, the order of i and j is important, and the matrix is no longer necessarily symmetric. The entry a_{ij} is set to 1 only when $x_i \rightarrow x_j$ is an arc in the directed graph (but not if only $x_j \rightarrow x_i$ is). For the directed graph above, the adjacency matrix is

```
0 0 1 1 1 1 0 0 0
1 0 1 0 0 0 0 0 0
0 0 0 0 0 0 0 0 0
0 0 0 0 0 0 0 0 0
0 0 0 0 0 0 0 0 0
0 0 0 0 1 0 0 0 0
1 0 0 0 0 0 0 0 0
0 0 0 0 0 0 0 0 0
0 0 0 0 0 0 0 0 1
```

The out—degree of a point can be found by totaling the row corresponding to that point. Likewise, the in—degree of a point is the sum of the entries of the column corresponding to that point.

A flow chart is a good example of a directed graph, since the direction of the arrows from one symbol to another is very important. Other examples of directed graphs will come up shortly when we apply some of these concepts to practical problems.

APPLICATIONS OF GRAPHS

We turn now to several examples that illustrate how graphs can be used to structure and solve a variety of important problems.

MINIMAL COST SPANNING TREES

The computer communications network problem described in Chapter 0 is a typical instance of how graphs are used to express complex problems. Recall that the communications network was expressed as a connected, undirected graph. Associated with each edge of the graph was a nonnegative cost. The problem involved finding a way to route a message from one point to another while minimizing the cost. The total cost of routing a message was computed by summing the costs on the edges in the routing path. Let us see how this general type of problem can be solved using some of the properties of graphs explored earlier in this chapter.

Suppose we have a connected, nondirected graph. Coupled with the graph, we have a list of costs, one assigned to each edge. The points of this graph can represent all sorts of things: building locations, computer sites, cities, and so on. The edges of the graph represent potential connections among the locations. The cost assigned to an edge can represent the cost of actually implementing the connection. The problem is to find a tree that joins all of the points of the graph and at the same time minimizes the cost of connecting the entire set of locations. For example, visualize the points of the graph as buildings on a university campus. The edges may represent the possible locations of trenches to be dug to lay utility cable. Your goal is to connect all of the buildings and to do so at the lowest cost. Clearly, multiple paths from one point to another increase the cost, so there will be exactly one path between any two points. In terms of what we have learned before, we are looking for a spanning tree of the graph. More importantly, since there can be more than one spanning tree for a given graph, we are searching for the spanning tree with the lowest cost. Hence, this type of problem is known as a *minimal cost spanning tree* problem.

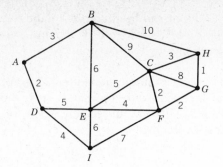

Figure 7.49

Let us work through an example to see how this type of problem is solved. Suppose the graph in Figure 7.49 is representative of the problem we are trying to solve. We will use an algorithm that examines the points in the manner described below, marking those which are to be included in the minimal cost spanning tree. The algorithm works in the following way:

1. Unmark all points in the graph.
2. Begin with any point in the graph; mark it. It will be the first point in our spanning tree.
3. Examine all unmarked points and find an unmarked point which has an edge of least cost connecting it to a marked point. (There may be more than one such point.) Mark this point, and add it and the edge connecting it to the spanning tree.
4. Repeat step 3 until the set of unmarked points is empty.

As we pointed out, the spanning tree obtained by using this algorithm may not be unique. The multiplicity of spanning trees results from the choices available to us in step 3 of the algorithm. However, the choice for the first point will not affect the optimality of the spanning tree.

Let us apply the algorithm to the graph in Figure 7.49 to illustrate its use.

1. Begin with point E. (E was chosen arbitrarily.)
2. The least expensive connection to an unmarked point allows us to add point F to our tree.
3. The least expensive connection to the set of points { E, F } is C (from F). Note that the cost to G is also minimal.
4. The least expensive connection to the set { E, F, C } is G (from F).
5. In the same manner, we add H (from G) to the set { C, E, F, G }.
6. Then, we add D (from E) to { C, E, F, G, H }.
7. We add A (from D) to { C, D, E, F, G, H }.

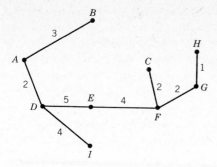

Figure 7.50

Figure 7.51

8. We add B (from A) to { A, C, D, E, F, G, H }.
9. Finally, we add I (from D) to { A, B, C, D, E, F, G, H }.

The resulting spanning tree is shown in Figure 7.50. If this were a computer communications network, this minimal cost spanning tree would represent the path on which there is exactly one connection between any two computers and for which the cost is minimized.

CRITICAL PATHS

The previous problem assigned costs to edges in an undirected graph. Suppose we assign instead nonnegative costs to the arcs of a directed graph. Manufacturers can represent their manufacturing processes or their construction tasks in such a manner. For instance, each point of the graph can represent a location in the manufacturing process. The arcs represent the tasks to be done, and the cost associated with each arc is the time it takes to complete the task. In this instance, there is a single starting point (beginning the process) and a single stopping point (finishing the process). Because the initiation of a particular task depends on the completion of the tasks that precede it, no task can be initiated from a point until all arcs feeding into that point have been completed. You can think of the graph as representing an assembly line. You cannot assemble the next component along the assembly line until all of the components for the current task have been received. Figure 7.51 represents a node for which three tasks must be completed before two others can begin.

We can think of the incoming arcs to a node as the prerequisites for that node. In this sense, the network describes the critical decision points in the process. We can view your computer science curriculum as a critical path network: certain courses and skills are required at any point before you can proceed to take the next course. In this sense, it is easy to see that you cannot go backward in such a network. In other words, you cannot take an advanced version of a course before you take the elementary ones. In general, this is a criterion for a critical path network: there can be no cycles in the network. Thus, there can be no set of arcs that lead out from a particular point and eventually return to that point.

Whether we view this sort of network as a curriculum or as a manufacturing scheme, several questions about the process times arise.

1. What is the smallest amount of time in which the entire process can be completed? In other words, what is the smallest amount of time until all the incoming tasks to the terminal point are done?
2. For each point in the process, what is the earliest time at which we can start all the tasks leaving that point? What is the latest time at which we can start all these tasks without slowing down or delaying completion of the entire process?

The difference between the two times defined in the second question is called the **float time**. A positive float time indicates that there is some leeway in the schedule. Limited delays at a point which has a positive float time may not slow down the entire process. A point with a float time of zero is a **critical node** in the process, since any delay at this point will delay the completion of the entire process. Special attention must be paid to those points with little or no float time. For example, noticing that certain points in a process have little or no float may encourage a manufacturer to train extra help to perform those tasks involved with the critical nodes.

A path from the starting point to the stopping point and including all the critical nodes is called a **critical path**. Such a path shows us which part of the process is most critical to the timing of the entire process. How do we calculate such a path? There is an algorithm for doing this.

1. At each point, there is an earliest time for each of the incoming tasks; this is the earliest time at which this particular task can be completed. For each point, we can take the maximum of all of these earliest times to calculate the earliest time at which all of the tasks coming into that point will be done. For example, if the point D in Figure 7.52 has tasks feeding it from points A, B, and C, we see that 35 is the earliest time (measured from the starting point) at which all of the tasks at A will be done and ready to start the task denoted by the arc from A to D. Likewise, 20 is the earliest time at which B will be done and ready to begin the task represented by arc $B-D$. Finally, 31 is the earliest time at which task $C-D$ can begin.

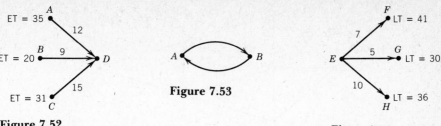

Figure 7.52

Figure 7.53

Figure 7.54

To compute the maximum earliest time for point D, we calculate

$$ET(D) = \max\,(35 + 12,\ 20 + 9,\ 31 + 15)$$

There can be no circularity in this digraph; that is, there can be no configuration of the form shown in Figure 7.53. Such a situation would say that B cannot be completed until A is completed, and also that A cannot be completed until B is. Note, too, that the earliest time at the starting point is 0.

2. Set the optimal result for the network to be the earliest time at the terminal point of the graph.
3. For each point of the graph, calculate the latest time for the point, that is, the latest time at which the point can be left (to begin the next tasks) without delaying the entire process. To calculate these latest times, work backward from the terminal point. The latest time for the terminal point is the earliest time (set in step 2). In the other cases, we minimize the times over all of the outgoing tasks. For example, if points E, F, G, and H are related as in Figure 7.54, then we calculate the latest time for E as

$$LT(E) = \min\,(41 - 7,\ 30 - 5,\ 36 - 10) = 25$$

This says that the product must be ready to leave point E at time 25. When this process terminates, the starting point should have a latest time of 0. Note that for any point x, the difference

$$LT(x) - ET(x) = \text{FLOAT time for } x$$

4. Finally, the critical path is formed by beginning at the starting point and moving to the terminal point by following the points of zero float. (How do we know that such points will exist?)

Let us work through an example to see how this works. Suppose we are given the directed graph of Figure 7.55 with the time constraints as noted.

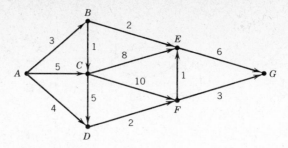

Figure 7.55

A is the starting point in this digraph, and G is the stopping point. First, we calculate the earliest times for each point to get

$ET(A) = 0$
$ET(B) = \max(0 + 3) = 3$
$ET(C) = \max(0 + 5, 3 + 1) = 5$
$ET(D) = \max(0 + 4, 5 + 5) = 10$
$ET(F) = \max(5 + 10, 10 + 2)$
$ET(E) = \max(3 + 2, 5 + 8, 15 + 1) = 16$
$ET(G) = \max(16 + 6, 15 + 3) = 22$

Next, we use the earliest time for G as the latest time for G to work backward and find the latest times for all the other points.

$LT(G) = 22$
$LT(E) = \min(22 - 6) = 16$
$LT(F) = \min(22 - 3, 16 - 1) = 15$
$LT(D) = \min(15 - 2) = 13$
$LT(C) = \min(13 - 5, 15 - 10, 16 - 8) = 5$
$LT(B) = \min(5 - 1, 16 - 2) = 4$
$LT(A) = \min(4 - 3, 5 - 5, 13 - 4) = 0$

Now that we have the earliest and latest times for each point, we compute the float:

$FLOAT(A) = 0$
$FLOAT(B) = 4 - 3 = 1$
$FLOAT(C) = 5 - 5 = 0$
$FLOAT(D) = 13 - 10 = 3$
$FLOAT(E) = 16 - 16 = 0$
$FLOAT(F) = 15 - 15 = 0$
$FLOAT(G) = 22 - 22 = 0$

The critical path in this graph is therefore the path that begins at A and travels to C, F, E, and finally to G.

We can formalize this algorithm to make it more suitable for implementation on a computer. Our directed graph is now viewed as a set P of points $\{ p_1, p_2, ..., p_n \}$, a set E of arcs $\{ e_1, e_2, ..., e_m \}$, and a set of edge costs $C = \{ c_1, c_2, ..., c_m \}$. We label the starting point p_s and terminal point p_t. We want to create time sets $A = \{ a_1, a_2, ..., a_n \}$ and $B = \{ b_1, b_2, ..., b_n \}$ where a_i represents the earliest time that we can leave point p_i and b_i the latest time we can leave p_i without delaying the entire process. Let S be the set of points for which we have calculated the value of a_i. Initially, then, we have only p_s in S and a_s equal to 0. Let T be the set of points for which b_i has been calculated. We will use T only after we have calculated a_i for all points. Initially, T will contain only p_t and $b_t = a_t$. The algorithm works as follows.

1. Define set S' to be the set of points which are in P but not yet in S.
2. If S' is empty, proceed to step 3. Otherwise, find a point in S' such that all incoming arcs to that point come from points in S. Call that point p_i, and let the points in S which have arcs to p_i be labeled p_{i1}, $p_{i2}, ..., p_{ir}$. Let a_i be the maximum of the terms $a_{ik} + c_{ik}$ for k ranging from 1 to r. Add point p_i to S and go to step 2.
3. Since S' is empty, we know that p_t is in S. Thus, a_t is the earliest time that the entire process can be completed.
4. Work back from p_t. Define T' to be the set of points in P but not in T. If T' is empty, proceed to step 5. Otherwise, find a point p_i in T' such that all arcs leaving p_i go to points in T. Call these points p_{i1}, $p_{i2}, ..., p_{iw}$. Let b_i be the minimum of the the values $b_{ik} - c_{ik}$ for k ranging from 1 to w. Add p_i to T. Repeat step 5.
5. At this point, we should have a_s and b_s equal to zero. Define the float times for each point p_i as $f_i = b_i - a_i$. All points whose float f_i is equal to zero are critical.

SHORTEST PATHS

A critical path in a directed graph traces the path most critical to the deadlines in completing a process. The components of the process "travel" (in some sense) along all of the arcs of the digraph. Suppose we now have a graph or digraph in which there is a choice of which arcs or edges of the graph to travel. There are still costs associated with each arc or edge, but this time we want to find the minimal cost from one given point to another. This sort of graph may represent distances from one location to another along various routes, and we are trying to get from one point to the other

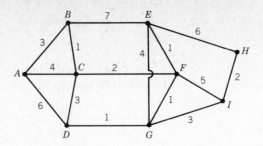

Figure 7.56

as cheaply as possible. The "costs" along each edge or arc could represent time, distance, money, or the like.

For example, let the network of Figure 7.56 represent the cost of traveling from point A to point H. It might represent interstate highways, for instance, and the cost associated with each edge is the cost in dollars for diesel fuel to cover the mileage along that segment of road. Or, the network may be a computer network. Each point is a computer, and the cost is the dollar value of a message sent from one computer to the next. We want to find the least costly path from A to H because it represents the least expensive connection.

The algorithm for finding the minimal cost path is similar to that for finding a minimal cost spanning tree. As with the spanning tree algorithm, we mark the points as we process the graph. This graph can be either undirected or directed; for a directed graph, the arc's direction governs permissible moves in the algorithm. The algorithm for finding the minimal cost path can be described as follows:

1. Unmark all of the points in the graph.
2. Mark the initial point (in this case, point A). There is a cost associated with each point relative to the initial point. The cost for the initial point is zero, since it costs nothing to go from the initial point to the initial point.
3. Find the least−cost unmarked point that can be reached from a marked point. The cost to get to a point is determined by the cost of the marked point at one end of the edge or arc plus the cost of traversing the next edge or arc. For example, if we are computing the cost from marked point X to unmarked point Y in Figure 7.57, we add the already computed cost of X to the cost of traversing the edge or arc from X to Y. Mark this next least−cost point.
4. Repeat the previous step until the terminal node (in the present example, H) is reached.

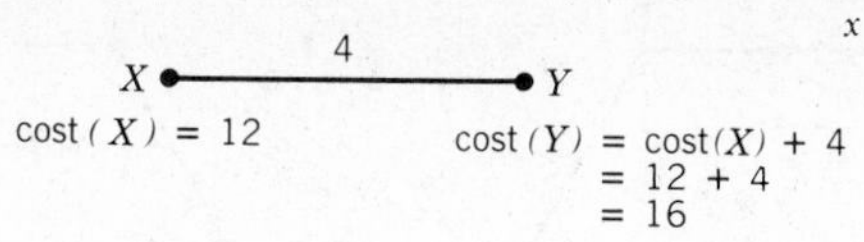

Figure 7.57

Let us follow the algorithm through to its conclusion for the graph shown in Figure 7.56. First, we set cost(A) to be zero, and we mark A. Next, we compute

$$\text{cost}(B) = 0 + 3 = 3$$
$$\text{cost}(C) = 0 + 4 = 4$$
$$\text{cost}(D) = 0 + 6 = 6$$

We choose B as the next point to be marked because it has the least cost from A. In the next iteration, we calculate

$$\text{cost}(C \text{ from } B) = 3 + 1 = 4$$
$$\text{cost}(E \text{ from } B) = 3 + 7 = 10$$

So we mark C and add it to our path. Continuing, we add D at a cost of 6 from C (we have a choice here), then F at a cost of 6, G at a cost of 7, E at a cost of 7, I at a cost of 7, and finally H at a cost of 12. Since H is the terminal point, we are done. Our least−cost path is thus

$$A \rightarrow B \rightarrow C \rightarrow D \rightarrow F \rightarrow G \rightarrow E \rightarrow I \rightarrow H$$

at a cost of $0 + 3 + 4 + 6 + 6 + 7 + 7 + 10 + 12 = 55$. Note that at C, we had a choice; there may have been other paths with the same minimal cost. Work through this example for some of the other choices to verify that the cost for the other paths generated by this algorithm is minimal.

We can state this algorithm formally to make its implementation on a computer easier to understand. First, we view the graph as three sets:

$$\text{a set of points } P = \{ \, p_1, p_2, \, ..., p_n \, \}$$
$$\text{a set of edges } E = \{ \, e_1, e_2, \, ..., e_m \, \}$$
$$\text{an edge cost function } C = \{ \, c_1, c_2, \, ..., c_m \, \}$$

The algorithm creates a point cost function

$$A = \{ \, a_1, a_2, \, ..., a_n \, \}$$

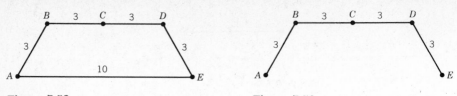

Figure 7.58   **Figure 7.59**

that represents the minimal cost of reaching each point. We call our start-ing point p_s and our target point p_t, where both are elements of the set P.

1. Let S be the set of points whose point cost functions have been calculated. Initially,

$$S = \{ p_s \} \text{ and } a_s = 0$$

2. If $P = S$ or if p_t is an element of S, then we are finished and can proceed to step 4.
3. Otherwise, we examine all edges connecting a point in S with a point in the set

$$P - S = \{ x \in P \mid x \text{ is not an element of } S \}$$

For all p_i in S, p_j in $P - S$, and edge e_k connecting p_i to $p_{j,}$ we calculate

$$\min(a_i + c_k)$$

In other words, we know the minimal cost to connect p_s to each point p_i in S. We find the point p_j in $P - S$ that can be reached next in the least expensive way. Go to step 2.
4. We now have p_t in the set S, so we have calculated the minimal cost to go from p_s to p_t.

In general, this method produces a minimal cost but not a minimal cost spanning tree. This is true even if we were to continue to calculate the costs of every point (so that $P = S$). We have found only the minimal cost of going from a given point to all other points. For example, for the graph in Figure 7.58, the minimal cost spanning tree is Figure 7.59. However, the minimal cost routes from A are shown in Figure 7.60.

MAXIMAL FLOW PROBLEMS

For a final example of how to apply graphs to practical situations, let us consider a directed graph. (The procedures described here will also work in

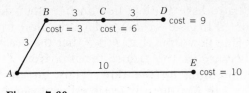

Figure 7.60

nondirected graphs, however.) Further, let us assume that the digraph with which we are working is connected. There are two special nodes in this digraph: one designated as a source, and another designated as the sink. The **source** is a point of in−degree zero and of positive out−degree. This means that all of the arcs incident with the source are leaving the source. Conversely, the **sink** is of zero out−degree and positive in−degree, so all the arcs incident with the sink are traveling into the sink. We can think of the digraph as describing the flow of something (information, water, computing, traffic, for instance) from the source to the sink. There is an unlimited supply of the flowing substance available at the source, and the costs associated with each arc of the digraph represent the maximum capacity of the arc. In other words, the cost tells us how many units of the flowing substance the arc can handle. The objective is to get the maximum number of units flowing from the source to the sink. Thus, such a problem is called a maximal flow problem.

As with the three previous examples, the algorithm which implements the solution to this problem labels the points of the digraph. A typical source−to−sink flow might be represented as in Figure 7.61. The **current flow** along an arc is the amount of flow at any stage of the calculation. An arc from point A to point B has a **feasible flow** if the current flow from A to B is less than the capacity of the arc. The feasible flow in such an arc is the remaining capacity. For instance, if we have a capacity of six in the arc from A to B, and if the current flow from A to B is 4, the feasible flow from A to B is two. This means that we can send two more units of flow along the arc from A to B.

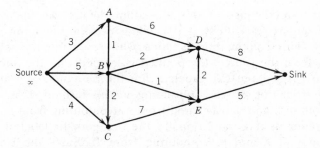

Figure 7.61

However, we also have feasible flow in the other direction, that is, in the direction from B to A, if we can reduce the flow from A to B. In this instance, the feasible flow from A to B is 2, but the feasible flow from B to A is 4. This is because we can reduce the flow from A to B by 4. Such a process is called **backing up the flow**. Thus, an arc from A to B is a **feasible arc** from A to B if there is feasible flow from A to B. At the same time, the arc can be feasible in the reverse direction. We will see how this concept is applied in the example that we work through.

First, we want to define an algorithm which will find a path from the source to the sink along which we can increase the flow from the source to the sink. We begin our algorithm by labeling the source with the infinity sign, ∞, to indicate that there is an infinite amount of flow available at the source. Let S be the set of labeled points. Thus, initially we have S containing only the source. Then, we proceed through the following steps:

1. Assign a current flow of 0 to each arc.
2. Unlabel all points except the source.
3. If the sink is in S, then we are done: proceed to step 4. Otherwise, let S' be the set of points of the graph that are not in S. For all points p_i in S and p_j in S', either there is an arc from p_i to p_j with feasible flow k in a forward direction, or there is an arc from p_j to p_i with feasible flow k in a reverse direction. Label p_j with the minimum of the feasible flow and the label on point p_i. Add point p_k to the set S. If there are no such points that can be labeled and we have not yet labeled the sink, go to step 5. Otherwise, repeat step 3.
4. The sink has been labeled with k. There is a single path from the source to the sink along which the labeling was done. Increment the flow along all arcs on this path by the amount k. This causes an increase in the current flow along forward arcs and a decrease in the current flow along reverse arcs. Go to step 1 to proceed with another iteration.
5. The flow can be increased no further in the network. The maximal flow is the flow leaving the source or entering the sink.

To illustrate the use of this algorithm, let us use the flow network pictured in Figure 7.61. First, A is labeled with the minimum of ∞ and 3, which is 3. This computation is done with respect to the source. Next, calculating from the source, the point B is labeled with the minimum of ∞ and 5, which is 5. Point C is labeled with the minimum of 4 and ∞, again calculating from the source. Continuing in this fashion, D is labeled with the minimum of 3 and 6, calculated from A. Computing from B, E is labeled with the minimum of 5 and 1. Finally, the sink gets the label of 3, since 3 is the minimum of 3 and 8, computing from D. Since the sink has been labeled with a 3, we increment all the flows by 3. Note that in the new

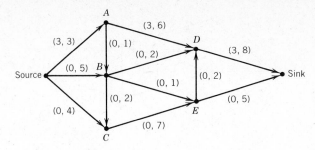

Figure 7.62

drawing of the graph in Figure 7.62, each arc has a pair of numbers associated with it. The first number is the current flow, and the second is the capacity.

Now we see that A cannot be labeled from the source, since there is no feasible flow. For B, we compute the minimum of ∞ and 5 from the source, and we give B a label of 5. C is labeled as the minimum of ∞ and 4 from the source, and D is labeled the minimum of 2 and 5 from B. E gets the label 1 because it is the minimum of 1 and 5 from B. Note that it is now possible to reduce the flow along the arc from A to D. Because of this, we label A with 2, which is the minimum of 2 and 3, calculating from D. Finally, the sink is labeled with the minimum of 2 and 5, calculating from D, where 5 is the remaining capacity on the arc from D to the sink. Thus, we increment the flow by 2, and the network becomes the graph of Figure 7.63.

For the next iteration, we have

B gets label $\min(\infty, 3) = 3$ from source
C gets label $\min(\infty, 4) = 4$ from source
E gets label $\min(3, 1) = 1$ from B
D gets label $\min(1, 2) = 1$ from E
sink gets label $\min(1, 5) = 1$ from E

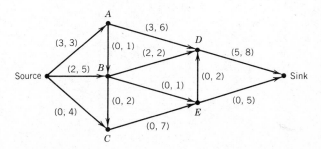

Figure 7.63

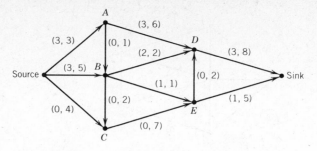

Figure 7.64

and the network can now be depicted as Figure 7.64.

Another iteration is necessary, and the results are

    B gets label 2 from the source
    C gets label 4 from the source
    E gets label 4 from C
    D gets label 2 from E
    sink gets label 4 from E

to yield the new network flows of Figure 7.65.

We can make another pass through step two of the algorithm, and we get

    B is labeled with 2 from the source
    C is labeled with 2 from B
    E is labeled 2 from C
    D is labeled 2 from E
    sink is labeled 2 from D

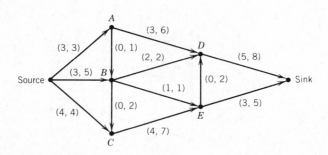

Figure 7.65

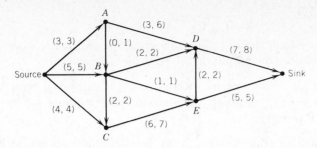

Figure 7.66

The network flow at this stage is shown in Figure 7.66.

Attempting to apply the next step in the algorithm, we see that nothing can be labeled from the source, and we cannot reach the sink. We are thus done, and the maximal flow is obtained by adding the current flow on the arcs leaving the source. Hence, the maximal flow is 12.

Note that in applying the steps of the algorithm, we must be prepared to back up flow along an arc whose current flow is positive. For example, in the network of Figure 7.67, we have labeled the sink from B by way of A. We can label B from the source with a 1. If we look only for remaining forward capacity, we get stuck, and we mistakenly think that the maximal flow is 1. However, we can label A from B, backing up the flow along the arc from A to B. This gives us the network flow of Figure 7.68, which doubles the flow from source to sink. When we back up flow along an arc, we are actually replacing flow. In this instance, the flow from B to the sink will come from the arc from the source to B. We must handle the flow that previously flowed along the arc from A to B; now, we send this flow to the sink.

As you have seen, graph theory can be an extremely useful tool in depicting and analyzing problems. The next chapter turns away from pictures and toward words: we examine the idea of a language.

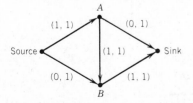

Figure 7.67

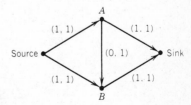

Figure 7.68

EXERCISES

1. This chapter defines when two graphs G and H are isomorphic. Show that this relation on graphs is an equivalence relation.

2. Draw all of the graphs without multiple edges on three points; four points; five points; six points. Is there an easy formula for the number of graphs on n points?

3. Reconstruct the graph G from the subgraphs shown in Figure 7.69.

4. Prove that if the length of a closed walk is odd, then the walk must contain a cycle.

5. Prove that the complete graph on n points is connected and is a regular graph of degree n.

6. Show that if a graph is connected, then any two longest paths in the graph must have a point in common.

7. Show that if a graph G is not connected, then its complement G' is connected.

8a. Draw all of the self–complementary graphs on n points for $n = 1$, 2, 3, 4, 5, 6.

b. Show that every self–complementary graph has $4\,n$ or $4\,n + 1$ points.

9. Show that the Ramsey numbers $R(m,\, n)$ satisfy the recurrence relation

$$R(m,\, n) \leq R(m-1,\, n) + R(m,\, n-1)$$

10. Let G be a finite connected graph with no loops or multiple edges. Prove that G has exactly one spanning subtree, or provide a graph as a counterexample.

11. Show that the complete graph on n points has exactly $C(n,\, 2)$ edges.

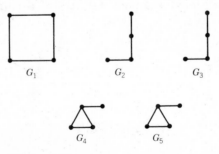

Figure 7.69

12. Prove that the graph in Figure 7.26 is isomorphic to the graph in Figure 7.27.

13. Prove Theorem 7.7 by induction.

14. Show that if G is a plane graph on p points in which the boundary of every region (except the infinite region) is a four−cycle, the number of edges of G must be $2p - 4$.

15a. Show that every planar graph on eight or fewer points has a planar complement.

b. Show that every planar graph on nine or more points has a non-planar complement.

16. Finish the proof of Theorem 7.11 by proving that every graph all of whose points are of even degree must be Eulerian.

17. Find the Eulerian trail in the graph of Figure 7.70.

18. Which of the graphs in Figure 7.71 are Eulerian? Which are Hamiltonian?

19. Prove that, for $n > 2$, if n is the length of the longest odd cycle of a graph G, then $\chi(G) \leq n + 1$.

20. For the path matrix D^k defined in this chapter, prove that if $D^k = D^{k+1}$ for some k, then $D^j = D^{j-1}$ must be the case for all j greater than k.

21. Let D be the adjacency matrix for a graph G. In this chapter, we found matrices D^k by using AND for $\times$ and OR for $+$ in matrix multiplica-

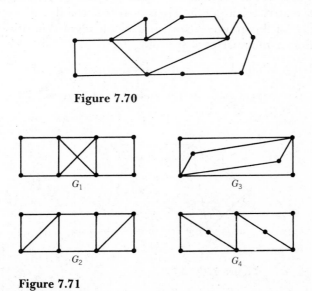

Figure 7.70

G_1

G_3

G_2

G_4

Figure 7.71

tion. Form a new set of matrices E^k from D by using matrix multiplication with $+$ substituted for $\times$ and MIN substituted for $+$. The operation MIN is defined to be

$$a \text{ MIN } b = \text{ the minimum of } a \text{ and } b$$

Show that this process terminates in a matrix whose $ij-$th entry represents the shortest path between points x_i and x_j.

22. PROGRAMMING PROBLEM: Write a program that reads the adjacency matrix of a graph and prints out a picture of the corresponding graph.

23. PROGRAMMING PROBLEM: Write a program that reads the incidence matrix of a graph and prints out a picture of the corresponding graph.

24. Which representation is easier to use in constructing a program to print out a picture of a graph: an adjacency matrix or an incidence matrix? Why?

25. PROGRAMMING PROBLEM: Write a program that reads the adjacency matrix of a directed graph and prints out a picture of the corresponding directed graph.

26. What does it mean when the adjacency matrix of a directed graph is symmetric?

27. PROGRAMMING PROBLEM: Write a program that implements the algorithm to produce a minimal cost spanning tree. This program reads in an adjacency matrix for a graph. Then, it reads in a list of costs associated with each edge of the graph. Finally, it uses the algorithm presented in this chapter to find a minimal cost spanning tree.

28. Find the minimal cost spanning tree in the graph of Figure 7.72.

29. Find the critical path in the network of Figure 7.73.

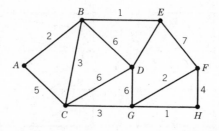

Figure 7.72

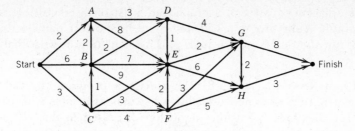

Figure 7.73

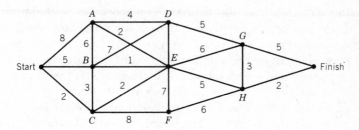

Figure 7.74

30. Calculate a shortest path in the graph of Figure 7.74. Is this the only shortest path? Is there a way to count the number of distinct shortest paths?

31. For the network shown in Figure 7.75, find the maximal flow.

32. PROGRAMMING PROBLEM: Write a program to implement the critical path algorithm. Your program will read in a description of a network, will find the critical path, and will print out the points on that path.

33. PROGRAMMING PROBLEM: Write a program that will accept as

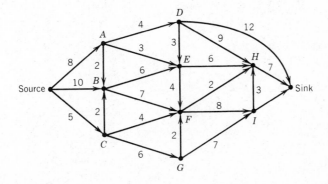

Figure 7.75

input the adjacency matrix of a graph and a list of distances for each edge of that graph. Your program will determine a shortest path in that graph and print out the points of that path plus a calculation of what that shortest distance is.

34. PROGRAMMING PROBLEM: Devise a program to read the adjacency matrix of a flow network plus a list that associates with each arc a capacity for the flow. Using the algorithm discussed in this chapter, your program should compute the maximal flow in the network and exhibit the flows along each of the arcs when that maximum amount is flowing.

35. PROGRAMMING PROBLEM: You work for a chemical company that is investigating the properties of hydrocarbon compounds. Hydrocarbons contain atoms of hydrogen, oxygen, carbon, and nitrogen. The **valence** of an element is the number of bonds that an element's atom can have to other elements. The valence of hydrogen is 1; the valence of carbon is 4; that of oxygen is 2; that of nitrogen is 3. Some typical compounds appear in Figure 7.76. The letters represent the atoms, and a line represents one bond. Bonds are bidirectional, so our illustrations use graphs

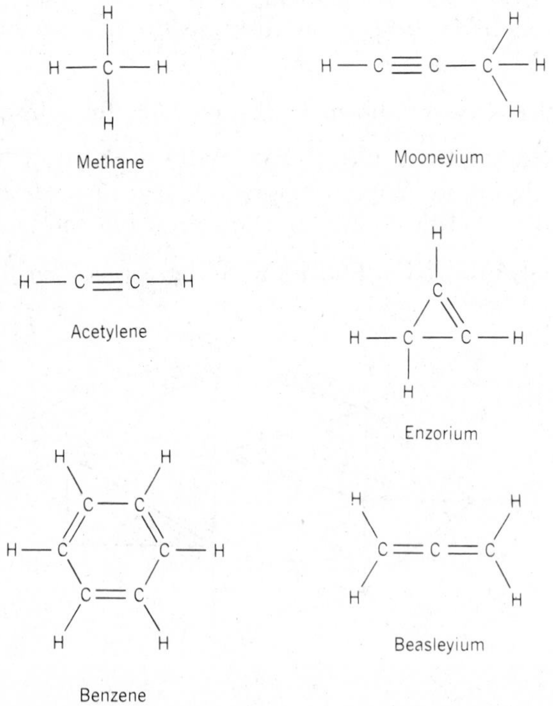

Methane

Mooneyium

Acetylene

Enzorium

Benzene

Beasleyium

Figure 7.76

rather than digraphs. Specifying the number and type of elements involved is not enough to distinguish among the compounds, because the bonds can be arranged in different patterns. Note that mooneyium and enzorium have three carbon and four hydrogen atoms each, but their bonds (and hence their properties) are different. We want to determine when one hydrocarbon compound is a subcompound of another. This problem is known as **graph matching**. Select a data structure which preserves the structure of a graph. Write a program that implements the data structure. Your output will indicate whether the two graphs are equivalent (i.e., the two hydrocarbons are equal), whether one is a subgraph of the other (i.e., one hydrocarbon is a subcompound of the other), or whether no such relation exists between the graphs. (Note: This is a difficult problem.)

CHAPTER 8

INTRODUCTION TO FORMAL LANGUAGES

In computer science, we speak freely of programming languages and their constructs. Just as we use natural languages to talk to each other, so, too, do we use a programming language to "talk to" or communicate with a computer. In this chapter, we examine more closely the concept of a programming language and what constitutes something we call a "language." Try to remember when you were learning a foreign language or what it is like when a child first learns to talk. You begin by learning a few words. Eventually, you string those words together to form sentence fragments. In the final stages, you learn how to generate well—formed sentences. At first, it is only an intuitive feeling that tells you when your sentence structure makes sense or is acceptable. Later on, however, the intuitive ideas are supplanted by a knowledge of the rules and grammatical structure of the language.

NATURAL AND FORMAL LANGUAGES

Such a language, that is, a language that we use for speaking or communicating, is known as a natural language. A programming language tends to be similar to natural language in that it aims to communicate and is made up of strings of "words" arranged according to a set of rules; however, such a language is much more rigid and much less expressive than natural language. The study of formal languages was begun by Chomsky in the 1950s. It relates and compares programming languages to natural languages. It determines what framework and methodology are required for something to be a language. This chapter applies the concepts we have explored in previous chapters to the principles involved in forming and using languages: natural languages, programming languages, and other theoretical entities that fit the description of a language.

Let us begin our investigation of language the way we did when we were children. Before learning to read a natural language, we must learn the alphabet for that language. Formally, an **alphabet** is a finite set E of symbols. A **string** is a finite sequence (possibly the empty sequence) of symbols, where each symbol is an element of an alphabet E. We can think of a nonempty string as an ordered $n-$tuple each element of which is in E. We say that a string is of **length** n if we can represent it as an ordered $n-$tuple. Let us call E^* the set of all strings of all lengths made up of symbols in the alphabet E. Sometimes E^* is referred to as the **set of strings over the alphabet** E. E^* is thus the set of all elements $(a_1, a_2, a_3, ..., a_n)$, where n is an integer and each a_i is in E. Note that E^* includes the string of length 0. This is called the **empty string**. To avoid consideration of the empty string in many instances, we define the set $E+$ to be the set of all strings of length 1 or greater over E. If we let e denote the empty or null string, then we can write

$$E^* = E+ \cup \{ e \}$$

Given the alphabet and the ability to form strings, how can we express the concept of a language? It is far simpler than you may think. A **language** L on the alphabet E is a subset of E^*. Although the strings are finite in length, a language may be finite or infinite. Let us look at a familiar example. Let E be the set

$\{ a, b, c, d, e, f, g, h, i, j, k, l, m, n, o, p, q, r, s, t, u,$

$v, w, x, y, z, A, B, C, D, E, F, G, H, I, J, K, L, M, N, O, P,$

$Q, R, S, T, U, V, W, X, Y, Z, 0, 1, 2, 3, 4, 5, 6, 7, 8, 9,$

$=, +, -, ;, :, ,, ., *, (,) \}$

We want to define a language on this alphabet and call it Pascal. How do we do that? We want Pascal to be considered the set of all valid Pascal programs, so how can we think of a Pascal program? A typical Pascal program is a string on E which is of a form similar to

```
PROGRAM SNORT(INPUT, OUTPUT);
    VAR ALPHA, BETA: INTEGER;
    BEGIN
    ...
```

and so on. Some strings on E represent valid Pascal programs, but others do not. It is important to realize that *any* set of strings over an alphabet is a language. We may put more restrictions on that set of strings to make it a particular kind or style of language, but the most general way of describing a language is simply as a set of strings. To restrict the more general language to the Pascal, we must have a method for defining Pascal.

GENERATIVE GRAMMARS

There are two useful methods for defining a language over an alphabet. The first is by specifying what is known as a *generative grammar*. You know from studying English that it is not enough to know words and phrases. The grammar of English specifies particular ways to put words or strings together. In a similar fashion, we can define a grammar for our formal language. We start with our alphabet E, and we look at another alphabet which has none of the symbols of E. We call this an **auxiliary alphabet,** and we call its elements **nonterminals**. Let N be an auxiliary alphabet for E. Then, N and E are disjoint, that is, $E \cap N = \phi$. A single element S of N is always designated to be the unique **starting symbol**.

Finally, we need a set of rules. Let P be a finite set of rules. We call these rules **productions,** and they must be written in a particular fashion. Let $(E \cup N)^* N (E \cup N)^*$ be the set whose elements are strings formed by taking a string in $(E \cup N)^*$, following it with a symbol from N, and ending with a string from $(E \cup N)^*$. A production is a mapping from the set $(E \cup N)^* N (E \cup N)^*$ to the set $(E \cup N)^*$. A **generative grammar** G is the $4-$tuple (E, P, N, S). We call the original alphabet E the set of **terminals** to distinguish it from the set of nonterminals N.

Let us look at how the generative grammar actually specifies the language. The productions take the form

$$xAy \rightarrow z$$

where A is a nonterminal, and x, y, and z are in the set $(E \cup N)^*$. Note that any of x, y, and z can be e, the null string. To generate a string over E^*, the generative grammar tells us to begin with the starting symbol S and apply productions until a string is reached that consists entirely of terminals. Thus, a production is actually a rule that specifies the replacement of one substring by another.

For example, let a and b be elements of E, and let A and B be non$-$terminals in N. Suppose the set of productions is the following:

     rule 1: $S \rightarrow aAB$
     rule 2: $S \rightarrow a$
     rule 3: $A \rightarrow bS$
     rule 4: $B \rightarrow b$

These productions generate the strings of a language by beginning with the starting symbol, S. For example, we can

1. Apply rule 1 to S to yield aAB.
2. Apply rule 3 to transform aAB to $abSB$.
3. Apply rule 4 to get $abSb$.
4. Apply rule 2 to get $abab$.

This is just one way that a series of rules can be applied. Clearly, there is an infinite number of ways we can apply these productions. Not only do we thus have a wide choice of production applications, but we also have a choice in the places in the string where we will apply the production. Therefore, a small number of productions can generate a large set of acceptable strings in our language.

The following is a more complicated set of productions over the same symbols.

rule 1: $S \to aAB$
rule 2: $S \to b$
rule 3: $A \to SS$
rule 4: $aA \to aab$
rule 5: $aB \to bb$
rule 6: $bB \to A$
rule 7: $B \to a$

Let us see how these productions generate the acceptable strings of our language. We can apply the productions to the starting symbol as follows:

1. Apply rule 1 to S to yield aAB.
2. Apply rule 7 to transform aAB to aAa.
3. Apply rule 3 to aAa, which results in $aSSa$.
4. Apply rule 1 resulting in $aaABSa$.
5. Applying rule 4 yields $aaabBSa$.
6. Apply rule 2 to get $aaabBba$.
7. Apply rule 7 to get $aaababa$.

We have seen how a series of productions results in a string. We can keep track of the rules used to generate the string; this is known as the **derivation** of the string. For the example above, we can denote the derivation of the string $aaababa$ by 1721427 to represent the rules involved.

FOUR CLASSES OF GRAMMARS

Languages differ in complexity. In some, the alphabets are small and simple. In others, not only are the alphabets much larger, but the productions are more complex and more numerous. We want to classify languages according to how complex they are. To do this, we break the set of languages into four types: *regular, context—free, context—sensitive,* and *unrestricted.* We give the same names to the grammars that specify these languages. The names are listed here in order of increasing complexity. If there is more than one type of grammar which can generate the same language, then we label that language by the simplest grammar that generates it. For example, a context—sensitive language can be generated by a context—sensitive grammar but not by any context—free or regular grammar.

Let us look at the four types of grammars in detail. Remember that we write productions in the form

$$xAy \rightarrow z$$

where A is a nonterminal and x, y and z are in $(E \cup N)$ *. A **regular grammar** has x and y equal to the null string, and z contains at most one element from N. Its productions look like

$$A \rightarrow a$$

or

$$A \rightarrow aB$$

where a is in E, A and B are in N. The production

$$S \rightarrow e$$

is allowed only if there is no production elsewhere with S on the right. There is an alternative way of generating regular languages. A **right—linear grammar** is one in which the productions are of the form

$$A \rightarrow xB$$
$$A \rightarrow x$$

and

$$A \rightarrow e$$

where x is an element of E^+ and A and B are elements of N. A right—linear grammar also generates regular languages.

A **context—free grammar** is one where the productions have x and y equal to the null string. The productions have the form

$$A \rightarrow x$$

where x is an element of $(E \cup N)^*$. Note that the string that results from the application of such a production cannot have fewer symbols in it than the initial string. Thus, a rule such as

$$AaBS \rightarrow aab$$

is not legal. The one exception to this is that a production of the form $A \rightarrow e$ is permitted. A typical production in a context—free grammar might be

$$A \rightarrow abAbB$$

where A and B are nonterminals while a and b are terminals.

The third class of grammars is called **context—sensitive**. Unlike context—free grammars, context—sensitive grammars have productions of the form

$$xAy \rightarrow z$$

where x, y, and z are elements of $(N \cup E)^*$. The production

$$S \rightarrow e$$

is also allowed provided that S never appears on the right—hand side of any rule. The string that results from the application of a production in a context—sensitive grammar must have at least as many symbols as the initial string. (The only exception is $S \rightarrow e$.) For instance, a production may look like

$$AaBS \rightarrow abbBBa$$

where A, B, and S are nonterminals and a and b are terminals. An alternative but equivalent definition says that a grammar is context—sensitive if the productions are of the form

$$uxAyv \rightarrow uzv$$

where u, x, y, v, and z are in $(N \cup E)^*$ and with z having at least as many symbols in it as xAy. This can be viewed as saying that xAy in the context of u and v may be replaced by z, and it is the reason we call such a grammar "context−sensitive."

The final class of grammars is called **unrestricted**. In this class, there are no restrictions placed on the productions.

RECOGNIZERS

Most modern programming languages are generated by grammars. In fact, the majority of these programming languages are mostly context−free. That is, the great majority of the productions satisfy the conditions for a context−free grammar, but there may be some productions which are context−sensitive. Why is this a concern? Why not consider all grammars to be unrestricted? The answer becomes apparent when it is time to write a compiler to parse the strings of the language. A grammar defines a language, but a parser must recognize when something is a valid "sentence" or program in the language. In this regard, a grammar is not particularly useful. It is certainly impractical to use the grammar to generate all valid programs in the language and then test the candidate program to see if it is one of the valid ones. For this reason, we look at a second way of defining a language.

A **recognizer** of a language is a procedure that examines a string and determines whether or not that string is a valid string of the language. In general, the simpler the type of grammar, the easier it is to produce a recognizer for it. Thus, it is desirable to keep a grammar as simple as possible. Just as we classified grammars, we can also classify recognizers.

1. A **finite automaton** or **finite state machine** recognizes regular languages.
2. A **pushdown automaton** recognizes context−free languages.
3. A **linear bounded automaton** recognizes context−sensitive languages.
4. A **Turing machine** recognizes languages generated by unrestricted grammars.

It is important to realize that there are languages which are not generated by any grammar. An infinite set of strings formed from $\{ 0, 1 \}^*$ where the 0s and 1s are random, for example, cannot be generated by any grammar. To visualize this, consider the random sampling of radioactive decay emissions. Assign a 0 if no emission is detected at a particular instant and a 1 if an emission is detected. The language consisting of all strings over the set { 0, 1 } contains as a proper subset the language generated by the radioactive emissions. However, since there is no pattern or regular routine, no grammar with a finite number of symbols and productions can generate this infinite set of random strings.

We will restrict ourselves to the four classes of grammars and recognizers defined above. Let us examine the types of recognizers in some detail.

REGULAR GRAMMARS REVISITED

As we saw before, regular grammars have productions of the form

$$A \rightarrow aB$$
$$S \rightarrow e$$

or

$$A \rightarrow b$$

for A, B, S in N and a, b in E. Note that a single nonterminal appears on the left and at most one nonterminal appears on the right. Let us look at some candidate productions and see whether or not they could be productions for a regular language:

1. $A \rightarrow e$: This is illegal. Only $S \rightarrow e$ is allowed.
2. $aB \rightarrow aA$: The left side of this production is illegal.
3. $B \rightarrow bA$: This production is fine.
4. $B \rightarrow Ab$: This one, though, is not permitted, because the nonterminal precedes the terminal on its right side.
5. $B \rightarrow AB$: This production is not legal because at most one nonterminal is allowed on the right side.
6. $A \rightarrow abbA$: Again, not permitted. Only one terminal is allowed on the right.
7. $A \rightarrow aS$: This is not permitted if $S \rightarrow e$ is included.

It is important to realize that in the process of generating a string x in E^*, we may arrive at a string that looks like

$$ababbabA$$

but can never reach a string looking like

abbAbabb

There can be only one nonterminal present, and it must be on the right—hand end. Because of this, we say that a regular language string has only one **point of generation**. This means, for example, that no rules of the form

$aA \to x$

(where x is in E) could be allowed. Thus, once we have generated a string *ababbaA*, we can add symbols at the right end by using another production ($A \to bB$, for instance), but the symbols *ababba* will always remain as the leftmost six symbols.

Remember, too, that a grammar has only a finite number of productions and finite alphabets E and N. In a regular grammar, we can view each nonterminal as a carrier of information or a representative of some situation. Let us see how. Suppose we have the grammar where $E = \{\ a\ \}$, N is the set $\{\ A, B, C, S\ \}$, and there are six productions. The starting point can lead to one of two places:

$S \to e$

or

$S \to aA$

We use a shorthand to denote the two consequences:

$S \to e \mid aA$

The vertical bar $\mid$ can be read as an *or*. The remaining productions, then, are

$A \to aB$
$B \to aC \mid a$
$C \to aA$

What does the language generated by this grammar look like? A careful look at the productions tells us that the strings of this grammar are strings of *a*s where there are $3n$ *a*s in each string (where n is a nonnegative integer). Thus, strings of the form

a

or

> *aa*

or

> *aaaa*

cannot be generated by this grammar. The "information" carried by the nonterminals can be viewed in the following manner:

> S says that 0 *a*s have been produced so far.
>
> A says that $3n + 1$ *a*s have been produced so far (n any nonnegative integer).
>
> B says that $3n + 2$ *a*s have been produced so far.
>
> C says that $3n$ *a*s have been produced so far.

Hence, applying the production

> $C \rightarrow aA$

means that we are always going from $3n$ *a*s to $3n + 1$ *a*s. Applying the rule

> $B \rightarrow a$

we always have $3n + 2$ *a*s before applying the production and $3n + 3$ (which is $3(n+ 1)$) after, at which point no more productions can be applied.

Let $L(G)$ be the language generated by this grammar. We shall prove formally that $L(G)$ contains exactly the set of strings of $3n$ *a*s. Thus, we want to prove that

$$L(G) = \{ x \mid x \in \{ a \}^* \text{ and } \mid x \mid = 3n \text{ for some } n \geq 0 \}$$

First, we show that any string of $3n$ *a*s for $n \geq 0$ is in $L(G)$. We do this by using induction on n.

Basis: The string of length 0 ($= 3 \times 0$) is in $L(G)$ since

> $S \rightarrow e$

is a production. Thus, for $n = 0$, a string of length $3n$ is in the language $L(G)$.

Induction step: Assume that a^{3n} is in $L(G)$. We want to show that $a^{3(n+1)} = a^{3n+3}$ is in $L(G)$. We note that the string a^3 is in $L(G)$

since

$$S \to aA \to aaB \to aaa$$

is a derivation. Thus, we can assume that n is at least 1. The last production in the derivation of a^{3n} must have been

$$B \to a$$

This means that we must have had the string $a^{3n-1} B$ at this point. Instead of applying the production

$$B \to a$$

we apply the productions

$$
\begin{aligned}
a^{3n-1} B &\to a^{3n-1} aC \\
&\to a^{3n-1} aaA \\
&\to a^{3n-1} aaaB \\
&\to a^{3n-1} aaaa = a^{3n+3}.
\end{aligned}
$$

Thus, any string of as of length $3n$ can be generated by $L(G)$.

Next, we show that if a string is in $L(G)$, then it must consist of $3n$ as for some n. Again, we use induction on n. First, we note that the only way to obtain a terminal string is by using the production

$$S \to e$$

(which is a string of length 0) or the production

$$B \to a$$

Thus, to get to this point, the string must be of the form $a^{3n-1} B$.

Basis: Suppose we have a string of the form $a^k B$, where $0 \leq k < 3 = 3 \times 1$.

Then, we must have used the production

$$S \rightarrow aA$$

and then applied the production

$$A \rightarrow aB$$

to yield $a^2 B$. If we then apply a production to B, we must get either $a^3 C$ or a^3. In either case, k is equal to 2.

Induction step: Assume that if we have a string of the form $a^k B$ in which $3n - 3 \leq k < 3n$ (for $n \geq 1$), then $k = 3n-1$. Let us examine a string of the form $a^j B$, where $3n \leq j < 3n + 3$. This production can have come only from $a^{j-1} A$. Similarly, a^{j-1} can have come only from $a^{j-2} S$ (which is illegal unless $j = 2$) or from $a^{j-2} C$. Likewise, $a^{j-2} C$ must have come from $a^{j-3} B$. We know that

$$3n \leq j < 3n + 3$$

so we must have

$$3n - 3 \leq j - 3 < 3n$$

By our inductive hypothesis, $j - 3 = 3n - 1$, or $j = 3n + 2$, and our proof is complete since from $a^j B$ we can get a^{j+1}.

Another example of a regular grammar produces strings of as and bs with an odd number of as and also an odd number of bs. Let E be the set $\{ a, b \}$ and N the set $\{ S, A, B, C \}$. The productions for this language are

$$S \rightarrow aA \mid bB$$
$$A \rightarrow aS \mid bC \mid b$$
$$B \rightarrow bS \mid aC \mid a$$
$$C \rightarrow aB \mid bA$$

The corresponding information carried by each nonterminal can be viewed as

S means that the number of as so far is even and that the number of bs so far is also even.

A means that so far we have an odd number of as and an even number of bs.

B means that so far we have an odd number of *b*s and an even number of *a*s.

C means that so far we have an odd number of *a*s and an odd number of *b*s.

For example, we can use these productions to generate the following sequence:

$$S \rightarrow aA$$
$$\rightarrow aaS$$
$$\rightarrow aaaA$$
$$\rightarrow aaabC$$
$$\rightarrow aaabaB$$
$$\rightarrow aaabaaC$$
$$\rightarrow aaabaabA$$
$$\rightarrow aaabaabb$$

Examine the productions to convince yourself that this grammar will produce any string with an odd number of *a*s and an odd number of *b*s. Also, verify that any string with an even number of either *a*s or *b*s cannot be produced.

It is not always evident from a general description of the acceptable strings of a language whether or not that language is regular. For instance, we may specify that we want to produce a grammar to generate strings of *a*s and *b*s where the number of *a*s is equal to the number of *b*s. This description is similar to the descriptions of the grammars we looked at above. Unfortunately, a regular grammar cannot produce strings of *a*s and *b*s with equal numbers of *a*s and *b*s.

Suppose that we could in fact do this with a regular grammar. This grammar must have a finite alphabet *N* and a finite set of productions. Let the number of nonterminals in *N* be *k*. Then our grammar must be able to generate the string composed of 2 *k* *a*s followed by 2 *k* *b* s:

$$aaa...aabbb...bb$$
$$\longleftarrow 2k \longrightarrow \longleftarrow 2k \longrightarrow$$

Because our hypothetical grammar is regular, at some point we would reach the point

$$aaa...aX$$
$$\longleftarrow 2k \longrightarrow$$

for some nonterminal *X* in *N*. At this point, we must be able to generate exactly $2k$ bs. Moreover, it will not be legal to generate only $2k-1$ bs or $2k+1$ bs. With only k nonterminals, this generation is impossible. Let us see why.

The **pigeon hole principle** says that if we try to place $n+1$ items in n receptacles or "pigeon holes," we must have at least one receptacle containing at least two items. We saw above that in the grammar under consideration, we must arrive at a string that looks like $a^i X$ for some nonterminal X. If we have only k nonterminals in our grammar and want to represent *all* strings of the form $a^i b^i$, then the pigeon hole principle tells us that there must be at least two strings of the form $a^j X$ and $a^w X$ for the same nonterminal X. From here, there is no way to guarantee that the X in the first string will produce exactly j bs while the X in the second string will produce exactly w bs. In other words, there is no way to use the nonterminal in the string to count the number of terminals as they are produced. Thus, while regular grammars can generate languages which can have relatively simple descriptions of the strings, they cannot "count" in the sense shown in this example.

FINITE AUTOMATA

With practice, one can develop a feel for regular languages. However, it is sometimes difficult for even a veteran of language theory to tell whether certain languages are regular. Let us investigate the properties of a recognizer for these languages. As stated earlier, a recognizer for a regular language is called a finite automaton. A finite automaton is a 5−tuple (E, N, S, F, P) defined as follows:

> E is a finite alphabet of terminals.
> N is a finite set of states.
> S is an element of N designated as the starting state.
> F is a subset of N whose elements are designated final states.
> P is a mapping from NXE to N.

The mapping P is called the **next−state function**. It can be viewed as a finite set of rules for moving from one state to the next. For each state R and each a in E, there is exactly one rule $P(R, a) = M$, where M is in N. Therefore, there is a transition for every possible state and for every element in E. It is impossible to move to a state and to be unable to move further; being "stuck" in a state will never happen.

A string x in E^* is **recognized** or **accepted** by the finite automaton A if the following sequence of events occurs. Let x be the string $a_1\ a_2\ a_3\dots\ a_n$, where each a_i is in E.

1. Begin in state S.
2. There is a unique rule that says that, given input string symbol a_1 when in state S, move to state B_1, where B_1 is an element of N. This can be written in shorthand as $(S,\ a_1) \rightarrow B_1$.
3. In general, we have $(B_j,\ a_{j+1}) \rightarrow B_{j+1}$ in N.
4. Finally, we have $(B_{n-1},\ a_n) \rightarrow B_n$ in N.
5. To be recognized, B_n must be a final state. In other words, B_n must be an element of F.

If B_n is not in F, the string is not recognized and is not in the language. Note, too, that if the string x is the empty string, we must have S in F to have x recognized. This says that the starting state must also be a final state of e to be recognized.

We can represent a finite automaton with a state diagram. For example, let E be the singleton set $\{\ a\ \}$. A double circle (as shown in Figure 8.1) indicates a final state, and an arrow with a letter indicates the transition to be made upon recognizing that letter. A finite automaton can be thought of as a machine which is fed a string of symbols. This machine begins at the left—hand side of the string, examines a symbol, moves to the next state, examines the next symbol, moves to the next state, and so on. When the last symbol is examined, the machine may or may not be in a final state. If it is in a final state, the string is accepted or recognized as being in the language; if it is not in a final state, the string is not in the language. What kinds of symbols does the machine above recognize? Experimentation will reveal that the sample machine recognizes strings containing exactly $3n$ as, where n is a nonnegative integer.

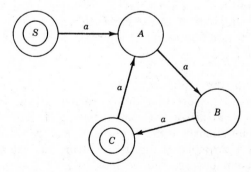

Figure 8.1

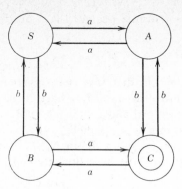

Figure 8.2

Let us look at another finite automaton. Its alphabet is $E = \{\ a,\ b\ \}$, and it can be represented as Figure 8.2. This automaton recognizes any string of as and bs where there are odd numbers of both as and bs.

DETERMINISTIC AND NONDETERMINISTIC FINITE STATE ACCEPTERS

The finite automata described here as recognizers of regular languages are also known as **deterministic finite state accepters (DFSAs)**. They are called deterministic because given a string x, the sequence of transitions that results is forced; that is, there is never any choice of moves from one state to the next. This is because the set of rules P is a function. There are also **nondeterministic finite state accepters (NFSAs)**. In these, P is not required to be a function. A string is accepted by an NFSA if there is a sequence of moves that concludes when the string is exhausted and the machine is in a final state. It is possible that other sequences of moves exhaust the string but result in a nonfinal state. However, as long as at least one such sequence ends in a final state, the string is accepted. It is for this reason that the accepter is nondeterministic: there is a choice of moves when leaving a given state.

For example, for the alphabet $E = \{\ a,\ b\ \}$, we might have the automaton of Figure 8.3. Note that there are two arrows labeled a which leave S, and there is no arrow with an a leaving A. The introduction of choice into the machine makes it much more difficult to decide whether the machine recognizes a particular string. Is the string $aabbaab$ accepted by this machine? The decision process can be lengthy.

Why, then, is it useful to have NFSAs? There are several good reasons. Suppose we have two DFSAs as pictured in Figure 8.4. DFSA 1 recognizes

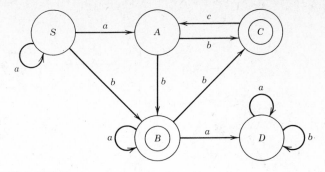

Figure 8.3

all strings of *a*s where the number of *a*s is odd. DFSA 2 accepts all strings of *a*s and *b*s that do not have two consecutive *a*s. (You can see from the diagram of DFSA 2 that two consecutive *a*s places you in state *B*, which is not a final state and from which there is no escape.) Suppose you want to recognize the union of these languages: all strings of *a*s and *b*s such that either the number of *a*s is odd or there are no two consecutive *a* s. Such a language is recognized easily by NFSA shown in Figure 8.5. As you can see, an immediate transition can be made into either DFSA 1 or DFSA 2.

The two DFSAs shown in Figure 8.4 have no arcs back to the starting state *S*. This makes the construction of an NFSA for the union quite easy. Consider instead the DFSAs depicted in Figures 8.2 and 8.3, where there are arcs back to *S*. To form an NFSA from the union of these, we begin by constructing the union of the DFSAs just as before. Then, for each arc from *S* to another state *K*, we add an identical arc from a new state *S'* to *K*. For each arc leading to *S*, we change the arc so that it leads instead to *S'*. Now, no arcs lead to *S* and the automaton recognizes the same language. To see why this is so, we list the sequence of states visited by the original automa-

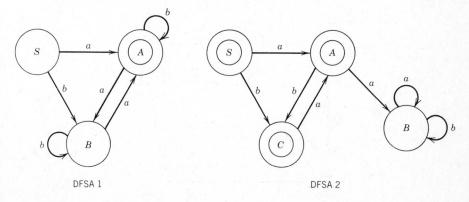

DFSA 1   DFSA 2

Figure 8.4

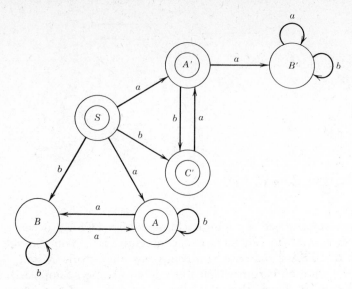

Figure 8.5

ton. Then, we modify the sequence by replacing S by S' for all occurrences of S except the first. For example, if the original sequence is

    SABSCSSDA

then the modified sequence is

    SABS'CS'S'DA

Figures 8.6 and 8.7 show how the automata of Figures 8.2 and 8.3 have been transformed.

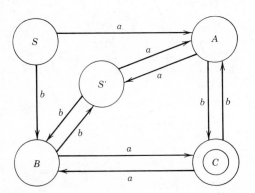

Figure 8.6

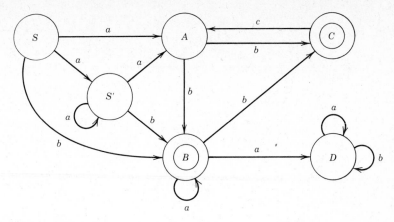

Figure 8.7

A second motivation for using an NFSA is related to the construction of a recognizer for a grammar. Suppose we have the regular grammar defined by the following productions:

$S \rightarrow aA \mid b$
$A \rightarrow aB \mid aA \mid a$
$B \rightarrow aC$
$C \rightarrow bS \mid aB \mid aA$

From these productions, the construction of an NFSA is relatively straightforward. For the rules $S \rightarrow aA \mid b$, we know that we need Figure 8.8 in our diagram. Adding the rules $A \rightarrow aB \mid aA \mid a$ gives us Figure 8.9. Note that for any rule of the form $G \rightarrow a$, we simply draw an arrow to the final state. Adding in the rest of the rules in the same way, we have Figure 8.10 as the result.

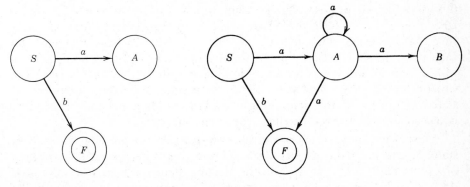

Figure 8.8   **Figure 8.9**

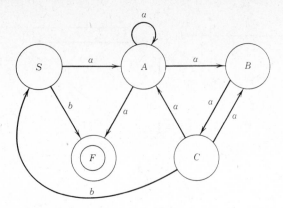

Figure 8.10

You can see how quickly an NFSA can be built from the rules of the grammar. In general, it is also easy to go from a finite state automaton (either DFSA or NFSA) to a grammar. To see how, suppose our automaton allows us to move from state A to state B over an arc labeled a. We include the rule $A \rightarrow aB$ in our grammar. If B is a final state, we also include the rule $A \rightarrow a$. If S is a final state with arcs leading into it, we modify the automaton as shown in the previous example by adding a new state S. Now, there are no arcs leading into S, and we include the rule $S \rightarrow e$ in our grammar. In this way, we retain the spirit of a regular grammar: $S \rightarrow e$ is allowed only if S does not appear as the right side of any production.

CONVERTING AN NFSA TO A DFSA

An NFSA is not as convenient to use as a DFSA, since we have to worry about making all sorts of choices. Happily, though, we can convert an NFSA into a DFSA which recognizes exactly the same language. To show you how this is done, we present an algorithm which transforms an NFSA into a DFSA. Suppose we have the NFSA depicted in Figure 8.11. First, we make sure that for every state, there is at least one arrow leaving that state for each possible alphabet symbol. In the example above, D has no exit arrow with a b on it, and A has no arrow leaving it with a b. To take care of this situation, we create a new state called a **trap** state. We have added the trap state in Figure 8.12 and denoted it by T. The trap is not a final state, but it is a state which, if ever reached, can never be left.

Next, we examine each of the states in turn. We begin with the starting state, S, and put it in our new DFSA. From this state, we create two exits, one for a and one for b. An a in the starting state sends us to a new state, AB, shown in Figure 8.13. This state AB is a combination of states A and B

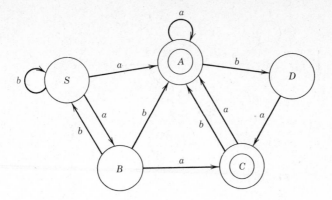

Figure 8.11

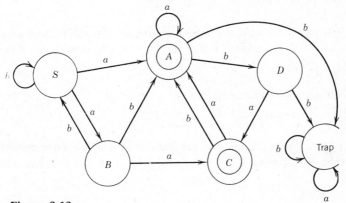

Figure 8.12

in the original machine; it is a final state because A was final in the original. Now we look at what states can be reached from states A and B in the original machine:

From A with an a we can reach A, and from B with an a we can reach C, so from AB with an a we can reach A and C.

From A with a b we can reach D and T, and from B with a b we can reach S and A, so from AB with a b we can reach D, T, S, and A.

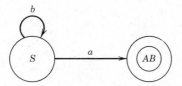

Figure 8.13

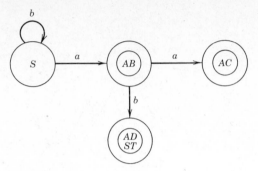

Figure 8.14

With this information, we generate two more states for our new machine. The machine is illustrated as Figure 8.14.

As before, we look at those states in the new machine and what we can reach from the corresponding states in the old machine. From state *AC*, we see that

> With an *a* we can reach state *A*
> With a *b* we can reach *D*, *T*, and *A*

We thus append new states to our new machine to correspond to this discovery; the result is Figure 8.15. Continuing, we see that from state *A* we can reach

> With an a: *A*
> With a *b*: *D* and *T*

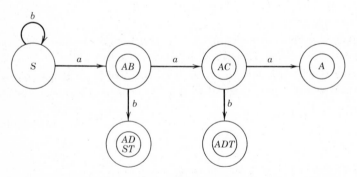

Figure 8.15

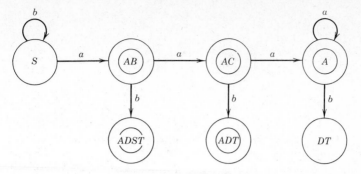

Figure 8.16

So, we now have Figure 8.16. Note that state *DT* is not final because neither *D* nor *T* is. At this point, it may seem that this process could go on forever. However, we are limited in generating the new states. We began with six states (including the trap state *T*), and we can count the number of possible combinations to examine:

$$C(6, 1) = 6 \text{ representing one state}$$
$$C(6, 2) = 15 \text{ representing two states}$$
$$C(6, 3) = 20 \text{ representing three states}$$
$$C(6, 4) = 15 \text{ representing four states}$$
$$C(6, 5) = 6 \text{ representing five states}$$
$$C(6, 6) = 1 \text{ representing six states}$$

Thus, there are at most $1 + 6 + 15 + 20 + 15 + 6 + 1 = 42$ possible states in the resulting machine. As you can see, this is a long process but one guaranteed to terminate, nevertheless.

As a final result, we have the machine of Figure 8.17. This is a DFSA which recognizes exactly the same set of strings as the original NFSA. How do we know? We can establish this by noting first that adding the trap state allowed us to recognize no new strings. Suppose that the original machine recognizes the string

$$a_1 \, a_2 \, a_3 \, a_4 \quad \ldots \, a_n$$

by going through the series of states

$$S_1 \, S_2 \, S_3 \, S_4 \ldots \, S_n$$

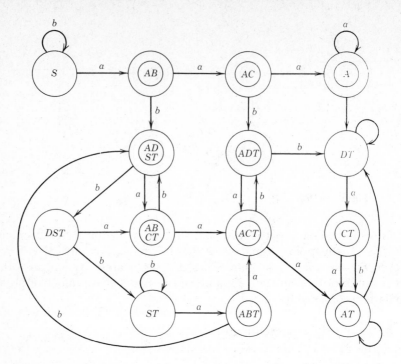

Figure 8.17

Figure 8.18   **Figure 8.19**

where S_1 is the starting state S and S_n is the final state F. This means that the original machine starts at S and, at the left—hand side of the string, recognizes a_1 and proceeds to S_2. From our construction process, the arrow in the DFSA with an a_1 on it led to a state that combined S_2. From S_2, we recognized a_3 and proceeded to S_3. Since we had a construct in the NFSA of the form shown in Figure 8.18, then the construction in the DFSA would have had a piece looking like Figure 8.19. Thus, every string recognized by the original machine is also recognized by the new machine. The demonstration that the strings recognized by the new DFSA are also recognized by the old NFSA is left as an exercise.

APPLICATIONS OF FINITE STATE MACHINES

Finite state machines are useful in many areas of computer science. For instance, a parity checker is a recognizer for a regular language. In this case, a $k-$bit parity checker recognizes a language consisting of strings of length $k + 1$ that are valid sequences. The machine accepts strings in which the parity bit is correct and rejects those in which the parity bit is incorrect.

Compilers often use finite automata to check for valid label or variable names. Only those labels or variable names which meet certain predefined criteria are recognized as correct.

Graph theory and finite automata have many parallels, since a finite automaton can be viewed as a directed graph. Many questions about the automaton can be rephrased and solved in graph—theoretic terms. For example, suppose a given finite state automaton has k states. Each state corresponds to a point in a directed graph. The final states can be designated on the digraph as labeled terminal points. Given the alphabet for the finite state automaton, we want to know whether the automaton accepts any strings. In other words, is the language recognized by this finite state automaton nonempty? In terms of the digraph, we can answer this question in the affirmative if we can find a walk from a starting point to any of the terminal points.

Once we have established that the language recognized is nonempty, we may want to know whether the number of strings recognized by the finite state automaton is finite or infinite. The answer can be found by counting the number of walks from a starting point in the digraph to the set of terminal points.

How do we know when to stop our testing? We may test all strings of length ranging from 0 to $k - 1$ and find that none is accepted. Should we continue and test all strings of length k? Equivalently, we may wonder what is the longest possible string that moves us to a different state each time? In the digraph, this means that we are seeking the longest possible path. Numbering the points (or states) as we move along such a path, when we reach point k, we have traversed $k - 1$ edges (or, equivalently, have examined $k - 1$ string symbols). Any path of length k or greater must thus have a loop or a cycle, because it must pass through at least one point at least two times. Hence, we can conclude that if no string of length $k - 1$ or less leads to a final state, then neither can a string of length k or greater.

Therefore, to determine whether any strings are accepted by a given finite state automaton, we need only to test all possible strings of length $k - 1$ or less. This knowledge allows us to construct an algorithm for testing finite state automata.

Suppose a string of length k or greater is accepted by our finite state automaton. As we saw above, this means that at least one state must be visited at least twice in the derivation of the string. A state visited twice is thus a point on a loop or cycle of the corresponding digraph. Since we can traverse this loop or cycle any number of times, we know that our finite state automaton will accept an infinite number of strings. Therefore, we must test all strings of length between k and $2k - 1$ to see whether any is accepted. If so, an infinite number of strings will be accepted.

The same questions can be considered by viewing the machine as a directed graph with a starting vertex and one or more final vertices. If there is a path from the starting vertex to a final vertex, then the machine accepts the string represented by that path. By checking all possible paths from the starting vertex to that final vertex, we can determine whether there is a path of maximum length. If a final state is on a cycle containing the starting vertex, then the strings can be of infinite length.

PUSHDOWN AUTOMATA

Let us turn to a consideration of recognizers for context—free languages. Remember that a context—free language has rules of the form

$$A \to x$$

where x is an element of $(N \cup E)^*$. Thus, rules such as

$$A \to BabA$$
$$A \to AB$$
$$A \to aba$$

are typical rules for a context—free language. Languages can be produced with context—free grammars that cannot be produced with regular grammars. For example, the grammar with productions

$$S \to aSb \mid ab$$

is context—free and produces the language $\{ a^k b^k \mid k > 0 \}$. As we saw earlier in this chapter, this is not a regular language.

A context—free language can have multiple points of generation, but the points of generation cannot be coordinated in any way. For example, if the generated string is

$$abAbabB$$

then the part of the resulting string that is generated from A cannot depend on the part that is generated by B. In other words, the generation of the language is independent of the context of the generators; thus, we call the language context—free.

How do we know that there are any languages which are not context—free? One way is to demonstrate that one is not. The language

$$L = \{ a^k b^k a^k \mid k > 0 \}$$

is not context—free. The way to prove this is by using a result known as the **theorem** *uvwxy*. This theorem is a useful tool in learning about formal languages. For more detail, we refer you to a text in the theory of formal languages.

As mentioned above, context—free languages are recognized by **push-down automata**. A pushdown automaton is much more complex than a finite automaton. As with a finite automaton, a pushdown automaton moves from left to right through the symbols in an input string as it moves from state to state. However, unlike a finite automaton, a pushdown automaton also has a memory. The memory is in the form of a structure called a **stack**. A stack is a data structure that resembles an office in—basket. The latest piece of data is placed on the top, so it is the first thing removed when it is time to examine the contents of the in—basket. Unlike the in—basket in an office, however, only the top item is visible in a stack. You cannot see how many items are in the stack, and you cannot reach into the middle. Another way of thinking of a stack is to liken it to the stack of dishes in the cafeteria. Often, the dishes are on a device which allows all but the top dish to sink into the device. When you want to place a new dish on the device, you must place it on the top; when you want to get a clean dish, you must choose the one on the top. Because the other dishes are below the surface of the device, you cannot tell by looking at the device how many dishes it contains. Placing an item on a stack is called **pushing the stack**. It works the same way as does placing or pushing a new dish onto the cafeteria device; when you are done, you see only the new dish on the top. Taking something off the stack is called **popping the stack**. In this instance, you see only the top of the stack, and you must take that item off.

A pushdown automaton, then, must have a stack alphabet as well as the regular alphabet. The action taken by the pushdown automaton depends on

1. The current state.
2. The input symbol.
3. The symbol on the top of the memory stack.

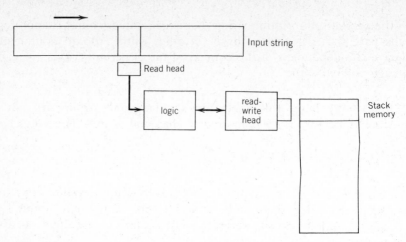

Figure 8.20

At any time, the stack can be pushed or popped. The action taken may or may not include erasure of the input symbol. Figure 8.20 shows how a pushdown automaton might look. For example, let us build a pushdown automaton to recognize the string xcx' over the alphabet $\{ a, b, c \}$, and with x in $\{ a, b \}$ *. The symbol x' means the reversal of the string x: if $x = abaab$, then $x' = baaba$. We begin in our starting state S. If we see an a or a b in the input string, we stay in state S and push the input symbol onto the stack. Then, we go to the next input symbol (or, erase the first input symbol in the input string, if you prefer to think of it that way). If we are in S and see a c in the input string, we move to state S_2 and erase c. Figure 8.21 shows us what happens. In state S_2, we have three choices:

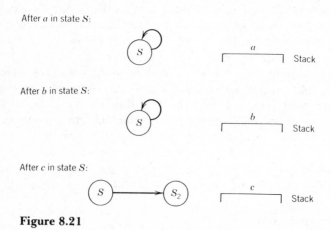

Figure 8.21

1. If the input symbol matches the symbol on the top of the stack, pop the stack and erase the input symbol.
2. If the input symbol does not match the symbol on the top of the stack, go to state S_3. The string is not accepted.
3. If we reach the end of the input string at the same time that the last stack symbol is popped, the string is accepted. Otherwise, the string is not accepted.

Work through this automaton to verify that it recognizes the language L. This language cannot be recognized by any finite automaton.

Given a context—free grammar, there are algorithms to generate a pushdown automaton that recognizes the context—free language. However, these algorithms are considerably more complex than those for producing finite automata from regular grammars. More complex still are the context—sensitive and unrestricted languages. An unrestricted language is recognized by a *Turing machine*, a device that operates on a magnetic tape that is infinite in both directions. The tape is divided into squares, each of which contains a symbol of the alphabet. A read—write head is positioned along the tape. At any moment, the read—write head can read a symbol on the tape, write the symbol, and move one square to the right or left. Unlike finite automata and pushdown acceptors (that can never go back to a previous input symbol), a Turing machine can return to examine characters that it has read before.

Context—sensitive languages are recognized by *linear bounded automata* (LBAs). We can think of a Turing machine as having an initial input tape, a read—write head that moves only in one direction, and an infinite auxiliary tape. The input string can be copied onto the auxiliary tape. An LBA is similar to this description of a Turing machine, but the auxiliary tape is of finite length. Typically, the auxiliary tape is no longer than the initial input string. The use of an auxiliary tape as a "memory" allows us more flexibility than a stack.

We examine Turing machines in greater depth in Chapter 9. If you are interested in learning more about these languages and algorithms, you should consult a textbook on formal languages.

DERIVATION TREES

Of what use are formal languages to computer scientists? One of the major applications is to the concept of a **derivation tree**. Such a tree is dependent

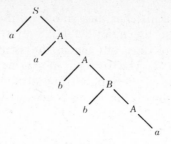

Figure 8.22

on a particular grammar, and it illustrates how a string in the language was produced. Let us look at some examples. The regular grammar

$$S \rightarrow aA \mid b$$
$$A \rightarrow bS \mid aA \mid bB \mid a$$
$$B \rightarrow bA$$

has the derivation tree for the string *aabba* shown in Figure 8.22. Each branch of the tree denotes which production was used at that step. We begin in state S, and we use the first production to move to aA. Thus, the branch

$$
\begin{array}{c}
S \\
/ \ \backslash \\
a \quad A
\end{array}
$$

shows us that S was replaced by aA. Likewise, the branch

$$
\begin{array}{c}
B \\
/ \ \backslash \\
b \quad A
\end{array}
$$

means that B was replaced by bA. The string produced is the set of leaves, or terminals, of the tree, read from left to right.

For the context—free grammar generated by these productions:

$$S \rightarrow AB \mid a$$
$$A \rightarrow Bab \mid b \mid aA$$
$$B \rightarrow bb \mid SBa$$

the corresponding derivation tree looks like Figure 8.23. The generated

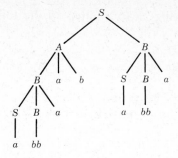

Figure 8.23

string is again denoted by the leaves, and it is read from left to right: *abbaababba*. Why are derivation trees important? Their importance lies in using compilers to parse strings such as

DELTA [NUM − 3]: = ALPHA [NUM + 1] − BETA * 2/(GAMMA + Y)

The easier it is to parse a string correctly, the faster and more efficient will be the compiler. Consequently, considerable effort is devoted to grammars which are geared toward ease of parsing.

AMBIGUITY

Sometimes, it is not clear how a string is generated from the rules of a grammar. A grammar is said to be **ambiguous** if there exists a string for which there are two different methods of generating that string. For example, the productions

$$S \rightarrow aA \mid aB$$
$$A \rightarrow b$$
$$B \rightarrow b$$

can generate the string *ab* in more than one way.

The order in which the productions are applied to generate a string is not important when deciding if a grammar is ambiguous. For instance, using the grammar

$$S \rightarrow AB$$
$$A \rightarrow a$$
$$B \rightarrow b$$

we can generate the string ab with either of the following two sequences:

$$S \rightarrow AB \rightarrow aB \rightarrow ab$$
$$S \rightarrow AB \rightarrow Ab \rightarrow ab$$

We call a **leftmost derivation** a derivation of a given string which always applies productions from left to right. Thus,

$$S \rightarrow AB \rightarrow aB \rightarrow ab$$

is a leftmost derivation, but

$$S \rightarrow AB \rightarrow Ab \rightarrow ab$$

is not. Ambiguity occurs when two different leftmost derivations produce the same result. A language is said to be ambiguous if every grammar which generates the language is ambiguous. There are no ambiguous regular languages. Why? Because every regular language is recognized by a finite automaton, and a finite automaton is deterministic. However, there are ambiguous context—free languages.

Some languages have more than one grammar to generate them, while others have no such grammar. Consider a language over the alphabet { 0, 1 } consisting of an infinite number of strings chosen at random. The strings can be viewed as binary representations of an infinite set of randomly generated positive integers. Since the strings are chosen randomly, no finite set of rules can generate the strings.

In this chapter, we delved into ideas relating to language. The exercises will give you practice in applying the ideas to problems with particular languages. The next chapter turns from language to what we in computer science often use a language for: computing.

EXERCISES

1. Construct a DFSA from the state transition diagram of Figure 8.24.

2. Produce an algorithm which, given a DFSA, will construct a regular grammar which defines the same language.

3. Let L_1 and L_2 be two regular languages.

a. Let G_1 and G_2 be two regular grammars generating L_1 and L_2, respectively. Define the concatenation of L_1 and L_2 as

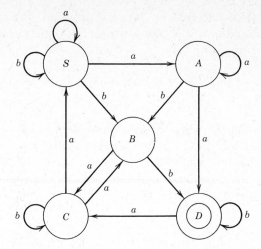

Figure 8.24

$$L_1 \bullet L_2 = \{\ xy\ |\ x \in L_1 \text{ and } y \in L_2\ \}$$

Produce an algorithm for constructing a regular grammar G_3 to generate L_1 $\bullet$ L_2. **Hint**: Instead of terminating with a string in L_1, how can we continue with the production of a string in L_2?

b. Let M_1 and M_2 be two DFSAs (finite automata) recognizing L_1 and L_2, respectively. Show how a DFSA M_3 is constructed which recognizes L_1 $\bullet$ L_2.

4. Let L be the language over $\{\ a, b\ \}$ consisting of equal numbers of as and bs. Construct a pushdown automaton to recognize L.

5. Let L be the language defined on $\{\ a, b\ \}$ consisting of all strings of as and bs which include the sequence $\llcorner abb$.

a. Produce a DFSA to recognize L.

b. Produce a regular grammar to generate L.

6. Characterize the language generated by the following grammar.

$S \rightarrow SSab\ |\ SSba\ |\ SaSb\ |\ SabS\ |\ SbSa\ |$
$SbaS\ |\ aSSb\ |\ aSbS\ |\ abSS\ |\ bSSa\ |$
$bSaS\ |\ baSS\ |\ ab\ |\ ba$

7. This chapter presents an algorithm for converting an NFSA into a DFSA. Show that the strings recognized by the new machine (i.e., the DFSA) are also recognized by the original machine (i.e., the NFSA).

8. PROGRAMMING PROBLEM: Write a program that will determine whether a finite automaton will accept a string. Let input record j represent string j in the following way:

Column 1 will contain an N for a nonfinal state or an F for a final state.

Columns 4 through 80 will contain a sequence of two—digit numbers, separated by spaces, which indicate the numbers of the states resulting from the input of an alphabetic symbol.

For example, the input

F 3 12 1 7 3

means that this is a final state, that input symbol a leads to state 3, that input symbol b leads to state 12, that input symbol c leads to state 1, and so on. In this way, reading in a series of cards defines the finite automaton. A card containing a dollar sign, $, as its first symbol indicates the end of the automaton specification. Your program will read in an input string of alphabetic characters and determine whether the string is accepted by this automaton.

9. PROGRAMMING PROBLEM:

a. Write a program that will determine whether an NFSA will accept a string. The specification of the NFSA is the same as that described in the previous problem.

b. Write a program to convert the NFSA to a DFSA.

10. PROGRAMMING PROBLEM:

a. Determine a method for describing a deterministic pushdown automaton so that its specification is suitable for computer input. This specification will require the creation of a stack. How else does this specification differ from those in the previous two problems?

b. Write a program, using your scheme from part (**a**), to read in the specification of a deterministic PDA and then read in strings for testing. Your program will determine whether the PDA will accept the strings.

CHAPTER 9

COMPUTABILITY

In studying computer science, you are often presented with well—defined problems. Indeed, you are usually told when a problem is solvable. Unfortunately, you do not always know the degree of difficulty of the solution, nor do you know the algorithm to use for the speediest solution. When you leave an academic environment, you will be asked to analyze a problem and determine the amount of time it will take to write a program to solve the problem. For each such problem, then, there are three major questions you must answer:

1. Is there a solution to this problem; that is, is the problem solvable? (The answer to this question will depend on your understanding of computation theory or the theory of computability.)
2. If the answer to question 1 is yes, how long will it take to produce a program to solve the problem? (The answer to this question is part of what you will learn in software engineering.)
3. If the answer to question 1 is yes, and you implement the solution according to a design developed as your answer to question 2, how efficiently will the program work with the data supplied to it? (The field of computer science known as algorithm analysis shows you how to answer this question.)

Because questions 2 and 3 depend on the answer to question 1, we will now introduce the theory of computability.

Frequently, when you look at the description of a problem, it is obvious that the solution exists. In other cases, it is equally obvious that no solution is possible. Sometimes, however, the answer to the question of solvability falls into a grey area between these two cases. Let us look at some examples to see why this is so.

Suppose you are asked to write a program which will calculate $n!$ for any value of n less than or equal to 1,000. At first, it seems clear that such a program can be written, since the formula for $n!$ is straightforward. However, on second thought, you realize that 1000! is an extremely large number; it will not fit into a full word or a double word on any existing computer. Thus, if a solution exists at all, it is far more complex than you may have originally thought. A third stab at the problem will reveal that you need not rely on the standard multiplication and addition techniques available for your computer. Possibly by making use of auxiliary memory (such as disks), you can work with numbers of effectively unbounded size. The answer to whether the problem is solvable, then, is yes, but the solution is not a trivial one. In fact, it is important to have more information about the problem. For instance, must the answer be exact? If not, an approximation technique (such as Stirling's formula) may be easier to implement and still be satisfying to the person who needs the answer.

On another occasion, you are asked to calculate the value $m + n$, given any two positive integers m and n less than 100. This problem is clearly solvable. Further, the accuracy of the answer does not make the problem more complex. However, in some instances accuracy may make a problem effectively unsolvable. For example, suppose you had to calculate the number of white blood cells in a particular person's body, count exactly the number of molecules in the universe, or print the decimal expansion of an irrational number. No program can calculate these numbers exactly. These problems are **intractable:** precision is needed and there may be algorithms to calculate the results, but there is no way of knowing just how accurate the results are. In other words, we may know that a correct algorithm exists, but we don't know which candidate algorithm is the right one.

What if you want to write a program to predict the outcome of college football games? Here, too, a yes—or—no answer as to the solvability of the problem cannot be given without further information about the accuracy and reliability required in the answer. Any student can write a program which will predict the results with a 95% certainty of not being off by more than 50 points for each team. (Think about that!) On the other hand, assuming no clairvoyance, it is not possible for a program to be 100% accurate to within zero points. Chance, including the initial coin toss and changing weather conditions, plays a part. What about the grey area between? When does this become a tractable problem? Suppose a program that is 90% accurate to within five points of the actual score is required. The solvability of this problem is not always evident.

You are probably familiar with chess tournaments pitting a computer program against a human opponent. Suppose you want to write a program whose performance is good enough to beat consistently the reigning human champion. The program must operate within the standard time limits for making the required number of moves per hour. Whether this problem is solvable within the time constraints is not an easy question to answer. Chess playing programs have been getting better and better, but they are not yet up to the world championship level. Improvements have been made both by increasing the sophistication of the programs and by incorporating technical advances that allow the machines to operate at higher speeds. Thus, the answer to whether the required program is tractable within the constraints may be no at present, but it might be yes in five years.

The previous examples give you some idea of the kinds of questions that must be answered and the kinds of problems that arise when trying to decide whether a problem is tractable. Whereas an intractable problem has a solution but we can't know what the correct algorithm is, an **unsolvable** problem is one for which no algorithm can ever be found that can produce a solution within a finite number of steps. The infinite loop problem introduced in Chapter 0 is an example of an unsolvable problem. Let us examine this problem in detail and see how to decide definitely whether the problem is solvable. In Chapter 0, we discussed writing a program that would read a program and its input as data and decide whether the program read contains any infinite loops. It certainly would be convenient for your Pascal compiler, for instance, to examine your source program and its input data and print out either

> This program and data have an infinite loop.

or

> This program and data are free of infinite loops.

Note that the first response is *not* the same as saying

> This program and data will not terminate within 400 minutes of CPU time.

This latter problem is solvable and depends on the properties of the machine on which the program runs. The processing time on a microcomputer will differ drastically from that of a Cray-1; something which runs for more than 400 minutes on a microcomputer may, in fact, terminate in far less time on the Cray. The description of the infinite loop problem is fairly simple, and your initial reaction may be that checking for infinite loops may be difficult but certainly feasible. However, if your company were to sign a contract to write such a Pascal compiler, you would be in serious trouble. Such a compiler cannot be written, and we can prove why. This type of proof reaches the heart of computability theory, and its result shows the importance of examining the solvability issue before tackling any problem with a program. As we will see, some problems are inherently unsolvable; solvability is independent of the computer on which the program is to be run.

Suppose that we could write a Pascal program P to check for infinite loops when presented with a Pascal program and its associated data. As described above, the program P will accept as input the Pascal program and data and will return as output either

No infinite loops

or

Has an infinite loop.

If we can write program P, we are equally capable of writing a more restricted program P' with the following specifications. P' receives as input only the Pascal program; it does not require the associated data. P' answers what may appear at first to be a strange question:

Given a Pascal program R, will R loop forever if we give R as input a copy of itself?

If you think about it, you will realize that any program which can meet the specifications of P must also be able to answer this question.

For the running of any program, we know that we must have one of three outcomes:

1. The program terminates abnormally; that is, the system detects an error and terminates the program.
2. The program terminates normally.
3. The program loops forever.

Thus, P' will print out whether R stops (either by normal or abnormal termination) or loops forever. If we can create Pascal program P, we can certainly create P'. P' simply specifies what the input data to R will be.

Now, we modify our program P' to create a new program, P''. Program P' prints out what program R does and then stops. If R loops forever, P'' will print this out, too, and stop. However, if R stops, then P'' will go into an infinite loop (perhaps by using a WHILE or a REPEAT-UNTIL loop). Thus, if program R stops, then P'' will loop forever.

The program P' may look something like this:

```
PROGRAM LOOPCHECK (INPUT, OUTPUT);
VAR
    RESULT: BOOLEAN;
    PROCEDURE TEST (RESULT: BOOLEAN);
```

```
        [Reads and tests a Pascal program:
        RESULT = T for no loops, F for infinite loops]
        END;
    BEGIN
        TEST (RESULT);
            IF RESULT
                THEN WRITELN ('NO INFINITE LOOPS')
                ELSE WRITELN ('LOOPS FOREVER')
    END
```

Program *P''* is identical except for the last six lines; they are changed to read as follows:

```
    BEGIN
        TEST (RESULT);
            IF RESULT
                THEN REPEAT-UNTIL 1 = 2 [generates infinite loop]
                ELSE WRITELN ('LOOPS FOREVER')
    END
```

Suppose we were to use *P''* as data for the program *P''*. What would happen? If *P''* (as *R*) stops when fed to *P''* (as the decision program), then *P''* would loop forever. On the other hand, if *P''* (as the data program *R*) loops forever when fed to *P''* (the decision program), then *P''* prints out the result and stops. As you can see, we have a contradiction. We are forced to conclude that there can be no such program *P*.

This problem is one of a class of problems known as undecidable problems. An **undecidable problem** is one for which there can be no algorithm that will calculate the result in any finite amount of time. A related problem is that of devising an algorithm that will determine whether any particular element is a member of a given set. We can reword the problem above so that it is equivalent to asking whether an algorithm exists which will determine whether a Pascal program belongs to the set of Pascal programs which stop when given themselves as input. As we have seen, this algorithm cannot exist.

Perhaps you are wondering whether the modification of the environment can have an effect on whether there is a solution to a problem. For example, since we cannot write a Pascal program to test for infinite loops, it may be the case that such a program can be written in a language such as ADA or Lisp. It is important to realize that, no matter what the language, the resultant program is translated by the compiler into machine language or interpreted by an interpreter written (usually) in machine language. Thus, any special properties of the starting language (such as the recursive nature of Pascal) are lost in the translation. Therefore, the solution cannot depend on a particular attribute or feature of a programming language. The only differences evidenced by changing from one language or computer to another are in terms of power and speed, not in terms of computability. Assuming unlimited auxiliary memory, computability is independent of computer or language.

How, then, do we approach the problem of computability? First, we must formalize our notion of what is a problem. We have seen above the expression of many problems that might confront us, but they all have something in common. A **problem** P is really just a set of questions. For instance, a sort routine can be viewed as a set of questions:

1. Are the next two elements in the proper order?
2. Which is the smaller of the two?

We say that a problem P is **solvable** or **decidable** if there exists an algorithm F which, when applied to a question q of P, produces the answer yes or no, depending on whether q has answer "yes" or "no." In other words, a problem is solvable if we can answer correctly each of the questions which comprise the problem. Thus, the problem of writing the Pascal program to determine whether a program has an infinite loop is an unsolvable problem.

Suppose, though, we can prove that a problem is solvable. Are we finished? Not really. A problem P is said to be **solved** if P is solvable and if we have indeed produced its solution. A **solution** in this case is an algorithm that will solve the problem. Obviously, finding a solution should follow deciding whether a problem can be solved in the first place. In this chapter, we examine in detail the idea of solvability. Several possibilities for algorithms will help us to decide on the solvability of a problem, and we will examine three of them in this chapter: Turing machines, recursive functions, and Markov algorithms.

TURING MACHINES

A Turing machine is a simple but very powerful way of picturing how a computer works. Any problem that can be solved on a computer can also be solved on a Turing machine. We can think of a Turing machine as being comprised of a handful of components:

An infinite tape

A read/write head

A finite instruction set

The infinite tape passes back and forth under the read/write head as commanded by the instruction set. Since the tape must be infinite, no one can construct a complete Turing machine. However, as we will see, the Turing machine is a useful tool for determining computability. Picture the tape of the Turing machine as consisting of squares on which symbols can be written, one symbol per square. A Turing machine has a finite set of states and operators over a finite alphabet E. The formal definition of a **Turing machine** requires a finite set of 5—tuples of the form

$$(S_1, a, S_2, b, M)$$

where

S_1 is the current state of the Turing machine.

a is the symbol under the read/write head.

S_2 is the next state: if we are in state S_1 and see an a on the tape, we move to state S_2.

b is the symbol to be written replacing a: a '-' means that nothing is to be written (a is unchanged).

M is the move instruction: R means move right one square; L means move left one square; $-$ means do not move.

Thus, each instruction corresponds to a 5—tuple. One state is designated as a starting state, and others can be designated as terminal states. (To what in Chapter 6 is this similar?) When a terminal state is reached, the Turing machine halts. An initial tape description shows what the tape looks like at the start, while an instantaneous tape description indicates what the tape looks like at any given time. The initial tape contains only a finite number of marked squares. In many instances, when a terminal state is reached, no 5—tuple is applicable; no further moves can be made. In fact, it is possible to have a Turing machine defined with no terminal states. In that case, we say that the machine halts when there is no 5—tuple which can be applied to the current state—symbol combination.

Numbers can be represented on a Turing machine tape as a series of 1s separated by 0s. For example, a 0 is represented by '1'; one by '11'; two by '111'; and, in general, the number k by a sequence of $k + 1$ 1s. The moves along the tape are uniquely determined for any situation (except for the terminal state), because there is a 5—tuple specifying exactly what action to take. Thus, a Turing machine is deterministic, just as the DFSAs were in the last chapter.

The method of representing a number as a sequence of 1s separated by 0s is the simplest way to define the Turing machine. Other methods require extra symbols. For example, if 0s and 1s were used to represent a number in its binary form, a third symbol would be needed to separate one binary number from the next. There is no reason that a Turing machine could not be defined in such a manner, but the classic approach minimizes the number of symbols and keeps the representation as simple as possible.

We can visualize a Turing machine tape as a book with an infinite number of pages all but a finite number of which are blank. Each page holds one symbol. You are the read—write head, and you have a pencil with an eraser. You can examine a symbol on a page, erase it, and write a symbol on the page. The finite instruction set is a finite collection of index cards; each card contains a configuration, a symbol, and a description of the action to take if, when in the given configuration, the symbol is encountered. If no card can be found that applies to the configuration, you stop. You can never turn over more than one page at a time, and there is an infinite number of pages both preceding and following the page under consideration.

To see how a Turing machine computes, let us look at several examples. The first Turing machine adds two numbers together. Initially, the tape looks like

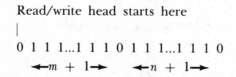

As you can see, the $m + 1$ 1s represent the number m, while the next $n + 1$ 1s represent the number n. The result, after processing by the Turing machine, should look like

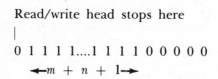

The Turing machine will move from the starting state, across the first number, and replace the 0 separating the two numbers with a 1. This means that (temporarily) we have $m + 1 + n + 1 + 1 = m + n + 3$ 1s. The machine continues across the second number and then begins a

journey to the left. It will erase one of the 1s and then change to a terminal state. This corresponds to the following sequence of 5—tuple instructions:

$(A, 0, B, -, R)$	Starting state: begin moving right
$(B, 1, B, -, R)$	Move across the first number
$(B, 0, C, 1, R)$	Move across to the second number
$(C, 1, C, -, R)$	Move across the second number
$(C, 0, D, -, L)$	Now move left
$(D, 1, E, 0, L)$	Erase a 1
$(E, 1, F, 0, L)$	Erase another 1 (we now have added one 1 and deleted two 1s)
$(F, 1, F, -, L)$	Move left
$(F, 0, G, -, -)$	Change to the terminal state (G is the terminal state)

Follow this sequence of 5—tuples carefully to see how the Turing machine adds the numbers. It will add any two numbers, including 0s.

The example above is fairly complex, considering that we are just adding two numbers. More complicated yet is using a Turing machine to read a number and write a copy of it. Given a number N, the starting position looks like

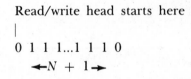

When the Turing machine is done, the tape will look like

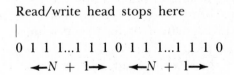

The Turing machine will operate as follows:

1. A marker is placed to the left of the original number, so that the result looks like

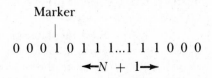

The Turing machine will be erasing (changing to 0s) the 1s in the original number, so the marker is used to "remember" the size of the original number.

2. A 1 is erased in the original number, and a 1 is added to a new number to the right of the original. This process is repeated until every 1 in the original has been copied in this fashion to the new number:

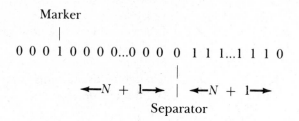

3. Using the marker as a stopping point, we begin with the position to the left of the separator and reconstruct the original number. When done, the Turing machine configuration is

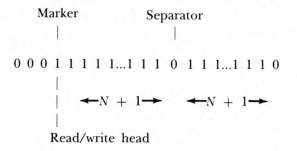

4. The read/write head is positioned over the 1 which has served as the marker. The Turing machine moves right to erase the marker and the extra 1 to its right. Our final state is thus

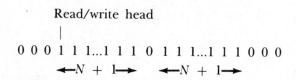

The sequence of instructions for the Turing machine is

$(A, 0, B, -, L)$	Starting state: move left
$(B, 0, C, 1, R)$	Place a marker on the tape
$(C, 0, D, -, R)$	Move to the beginning of the number

(D, 1, E, 0, R)	Erase a 1
(E, 1, E, −, R)	Move to the right to copy the 1
(E, 0, F, −, R)	Prepare to write a 1 in the copy
(F, 1, F, −, R)	Move across the copy
(F, 0, G, 1, L)	Insert the 1 in the copy
(G, 1, G, −, L)	Move left across the copy
(G, 0, H, −, L)	Now at the boundary between the two numbers
(H, 1, I, −, L)	Move left across the original number
(H, 0, M, 1, L)	If we had two consecutive zeros, the original number must now be erased
(I, 1, I, −, L)	Move left across the original number
(I, 0, D, −, R)	Prepare to erase another 1 in the original number
(M, 0, M, 1, L)	Replace the ones in the original number
(M, 1, T, 0, R)	We must have reached the marker placed here in the second step
(T, 1, U, 0, −)	Erase the other 1 placed as a marker (U is the halting state)

Let us illustrate how this works by copying the number 3 using this Turing machine. The arrow indicates the placement of the read/write head in each case.

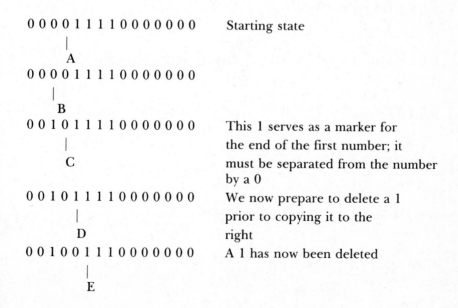

0 0 0 0 1 1 1 1 0 0 0 0 0 0 0
   |
   A
	Starting state

0 0 0 0 1 1 1 1 0 0 0 0 0 0 0
   |
   B	

0 0 1 0 1 1 1 1 0 0 0 0 0 0 0
     |
     C
	This 1 serves as a marker for the end of the first number; it must be separated from the number by a 0

0 0 1 0 1 1 1 1 0 0 0 0 0 0 0
       |
       D
	We now prepare to delete a 1 prior to copying it to the right

0 0 1 0 0 1 1 1 0 0 0 0 0 0 0
        |
        E
	A 1 has now been deleted

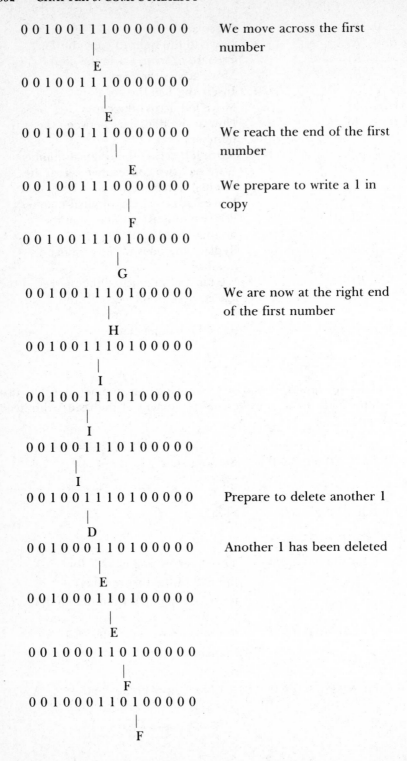

0 0 1 0 0 1 1 1 0 0 0 0 0 0 0   We move across the first
| number
E

0 0 1 0 0 1 1 1 0 0 0 0 0 0 0
|
E

0 0 1 0 0 1 1 1 0 0 0 0 0 0 0   We reach the end of the first
| number
E

0 0 1 0 0 1 1 1 0 0 0 0 0 0 0   We prepare to write a 1 in
| copy
F

0 0 1 0 0 1 1 1 0 1 0 0 0 0 0
|
G

0 0 1 0 0 1 1 1 0 1 0 0 0 0 0   We are now at the right end
| of the first number
H

0 0 1 0 0 1 1 1 0 1 0 0 0 0 0
|
I

0 0 1 0 0 1 1 1 0 1 0 0 0 0 0
|
I

0 0 1 0 0 1 1 1 0 1 0 0 0 0 0
|
I

0 0 1 0 0 1 1 1 0 1 0 0 0 0 0   Prepare to delete another 1
|
D

0 0 1 0 0 0 1 1 0 1 0 0 0 0 0   Another 1 has been deleted
|
E

0 0 1 0 0 0 1 1 0 1 0 0 0 0 0
|
E

0 0 1 0 0 0 1 1 0 1 0 0 0 0 0
|
F

0 0 1 0 0 0 1 1 0 1 0 0 0 0 0
|
F

```
0 0 1 0 0 0 1 1 0 1 0 0 0 0 0
              |
              F
0 0 1 0 0 0 1 1 0 1 1 0 0 0 0
             |
             G
0 0 1 0 0 0 1 1 0 1 1 0 0 0 0
            |
            G
0 0 1 0 0 0 1 1 0 1 1 0 0 0 0
           |
           H
0 0 1 0 0 0 1 1 0 1 1 0 0 0 0
         |
         I
0 0 1 0 0 0 1 1 0 1 1 0 0 0 0
        |
        I
0 0 1 0 0 0 1 1 0 1 1 0 0 0 0
        |
        D
0 0 1 0 0 0 0 1 0 1 1 0 0 0 0
         |
         E
0 0 1 0 0 0 0 1 0 1 1 0 0 0 0
          |
          E
0 0 1 0 0 0 0 1 0 1 1 0 0 0 0
           |
           F
0 0 1 0 0 0 0 1 0 1 1 0 0 0 0
         |
         F
0 0 1 0 0 0 0 1 0 1 1 0 0 0 0
          |
          F
0 0 1 0 0 0 0 1 0 1 1 1 0 0 0
          |
          G
0 0 1 0 0 0 0 1 0 1 1 1 0 0 0
         |
         G
```

0 0 1 0 0 0 0 1 0 1 1 1 0 0 0
|
G

0 0 1 0 0 0 0 1 0 1 1 1 0 0 0
|
H

0 0 1 0 0 0 0 1 0 1 1 1 0 0 0
|
I

0 0 1 0 0 0 0 1 0 1 1 1 0 0 0
|
D

0 0 1 0 0 0 0 0 0 1 1 1 0 0 0
|
E

0 0 1 0 0 0 0 0 0 1 1 1 0 0 0
|
E

0 0 1 0 0 0 0 0 0 1 1 1 0 0 0
|
F

0 0 1 0 0 0 0 0 0 1 1 1 0 0 0
|
F

0 0 1 0 0 0 0 0 0 1 1 1 0 0 0
|
F

0 0 1 0 0 0 0 0 0 1 1 1 1 0 0
|
G

0 0 1 0 0 0 0 0 0 1 1 1 1 0 0
|
G

0 0 1 0 0 0 0 0 0 1 1 1 1 0 0
|
G

0 0 1 0 0 0 0 0 0 1 1 1 1 0 0
|
G

0 0 1 0 0 0 0 0 0 1 1 1 1 0 0   Now we recreate the first
| number
H

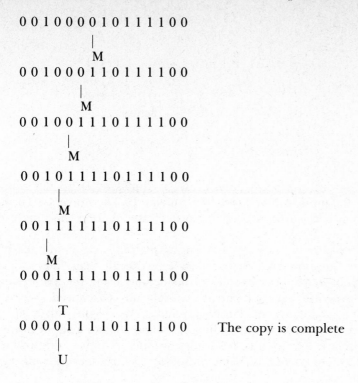

The copy is complete

Your first reaction may be that this is much ado about nothing. However, adding two numbers and copying numbers are fairly sophisticated activities for a machine which can only read 0s and 1s. Let us see what else Turing machines can do.

UNIVERSAL TURING MACHINES

A **universal Turing machine** is a Turing machine that emulates other Turing machines. Suppose, for instance, we have a Turing machine T and its initial tape description D. We can create from these a new initial tape description D' which includes a description of T on it. A 5−tuple such as

$$(3, 0, 5, -, R)$$

might appear on the tape as

State number

⟵⟶

80 1 1 1 1 0 1 0 1 1 1 1 1 0 1 1 1 0 1 1 1 0

| Next   Write   Move

Scanned   state   action

square
symbol

This universal Turing machine can use the 5—tuples in D' to determine how to process the input D to T ; thus, we emulate D. The universal Turing machine can then compute anything that a dedicated Turing machine can.

How can we determine exactly what Turing machines can compute? We can answer this question with a formal system. Recall that, in a formal system, we must have axioms and rules of deduction. Using the axioms and the rules, we generate new statements whose truth is assured. We can create something analogous for Turing machines. We begin with a set of Turing machines that perform simple functions, and we add a set of universal Turing machines that are given Turing machines as input and produce new Turing machines as output. This allows us to build new Turing machines of great complexity.

As an example, we will illustrate how Turing machines can be combined in this system. Suppose T_1 and T_2 are Turing machines which begin with a single number on a tape and produce a single number (not necessarily the same one) as a result. That is, each of T_1 and T_2 computes a unary function of its input. We define a third Turing machine T_3 in the following way. Turing machine T_1 starts just to the left of the number k on its input tape and produces a resulting number $T_1 (k)$. Likewise, Turing machine T_2 reads a tape and produces a result $T_2 (k)$ in the same fashion. To construct T_3, first we list all 5—tuples of T_1. Let E be the halting state of T_1. We then modify T_2 so that E becomes the starting state of T_2. In addition, we change the names of the states of T_2 to ensure that they do not conflict with the state names in T_1. This new, modified list of both T_1 's and T_2 's 5—tuples is the description of our machine T_3. T_3 operates by calculating $T_1 (k)$ and then continuing to produce $T_2 [T_1 (k)]$.

Can we use an analogous procedure to calculate a function

$$T_3 [T_1 (k_1), T_2 (k_2)]?$$

In this case, T_3 begins with two numbers on the tape, so the starting tape looks like

$$\xleftarrow{\hspace{0.5em}} k_1 \xrightarrow{\hspace{0.5em}} \qquad \xleftarrow{\hspace{0.5em}} k_2 \xrightarrow{\hspace{0.5em}}$$

...0 1 1 1...1 1 1 0 1 1 1 1...1 1 1 0...

There is one difficulty with this approach. T_1 is defined so that it assumes that there is only one number on the tape. As a result, the presence of k_2 can interfere with the working of T_1. One solution is to stop the normal working of T_1 when we encounter k_2. We may need to modify T_1 to allow it to sense interference from k_2. When we reach k_2, then, we stop temporarily and push k_2 one square to the right by changing the leftmost 1 in k_2 to a 0 and adding a 1 on the right. At this stage, we are likely to have a string that resembles

$$\xleftarrow{\hspace{0.5em}} T_1 \,(k_1) \xrightarrow{\hspace{0.5em}} \qquad\qquad \xleftarrow{\hspace{1em}} k_2 \xrightarrow{\hspace{0.5em}}$$

0 1 1 1...1 1 1 0 0...0 0 1 1 1...1 1 1 0...

|

Here, another intermediate routine must be added to move the tape reader over to a position just to the left of k_2 so that the calculation of $T_2 \,(k_2)$ can begin. If we find that $T_1 \,(k_1)$ interferes with the calculation of T_2, then we add instructions which push $T_1 \,(k_1)$ one square to the left by deleting the rightmost 1 and adding a 1 on the left. We thus would have

$$\xleftarrow{\hspace{0.5em}} T_1 \,(k_1) \xrightarrow{\hspace{0.5em}} \qquad\qquad \xleftarrow{\hspace{0.5em}} T_2 \,(k_2) \xrightarrow{\hspace{0.5em}}$$

0 1 1 1...1 1 1 0 0 0...0 0 0 1 1 1...1 1 1 0

|

Another intermediate step moves $T_2 \,(k_2)$ so that it is adjacent to $T_1 \,(k_1)$ and then moves the read/write head to the proper starting position:

$$\xleftarrow{\hspace{0.5em}} T_1 \,(k_1) \xrightarrow{\hspace{0.5em}} \qquad \xleftarrow{\hspace{0.5em}} T_2 \,(k_2) \xrightarrow{\hspace{0.5em}}$$

0 0 1 1 1...1 1 1 0 1 1 1...1 1 1 0

|

It is at this point that we can commence the calculation of $T_3 \,[\, T_1 \,(k_1), T_2 \,(k_2)]$

It is in this fashion that simple Turing machines can be combined to generate more complex ones. We end up with a formal system of Turing machines.

THE HALTING PROBLEM

Do all of the Turing machines in our formal system calculate functions and then stop? Unfortunately, the answer is no. Because a Turing machine operates with a set of 5—tuples and an initial tape description, it is possible that for some initial tape descriptions, the Turing machine would never halt. It is also possible for the Turing machine to get stuck. For example, suppose the initial tape description is not what it is supposed to be. The Turing machine may be in state S and expect to see a 0, and there is a 1 present instead. If there is no 5—tuple of the form $(S, 1, *, *, *)$, the Turing machine stops, just as a computer crashes in a similar situation. At other times, it is possible for a Turing machine to enter an infinite loop. Suppose, for instance, that the Turing machine contains the 5—tuples

$$(S, 0, S, -, R)$$

and

$$(S, 1, T, -, L)$$

and the read/write head is looking for a 1. This pair of instructions moves the head to the right, looking for a 1 on the tape. If there is no 1, the head will continue forever on the infinite tape.

Is it possible to examine the set of Turing machines and their initial tapes and decide whether they will stop or loop forever? (Does this problem sound familiar?) The answer is no, and the explanation is similar to the one used in explaining why we can't tell if a Pascal program will loop forever. Suppose we can build a Turing machine T which will take as input the set of Turing machines and their input tapes. Suppose further that this machine T stops after inserting the string 010 on the tape to indicate that the input machine halts or a 0110 to indicate that the input machine will loop forever. If we can build this machine T, then we can certainly build a universal Turing machine T' which accepts as its input a tape containing the description of a Turing machine Q; T' answers the question of whether Q halts when given Q as input. From T' we can produce a third Turing machine, T'', so that T'' loops if Q halts and halts if Q loops. Finally, we submit T'' to itself as input (which is permissable, since T'' is a Turing machine). This leads to the same contradictions we observed when we tried to find a Pascal program which checked for infinite loops.

The fact that such a Turing machine cannot be constructed is known as the **unsolvability of the halting problem**. What implications does this have for computing? For one thing, it again shows that there are unsolvable problems. We stated at the beginning of this chapter that anything that can be computed on a computer can also be computed using a Turing machine. Thus, the halting problem is an example of something for which we needn't bother to try to write a program, since no program will work. In this sense, the solution is uncomputable.

Turing machines allow us to represent the familiar aspects of computing (as we know it in our programming courses) as a formal system. This formal system uses simple components and a simple machine. It makes credible the notion that if a problem cannot be solved using a Turing machine, then it cannot be solved using any computer. Having a simple machine with which to work makes it relatively easy to establish proofs and theorems and to demonstrate the equivalence of other formal systems.

The Turing machine approach to computing is a very mechanical one because the computer is a mechanical object that scans a tape. There are many other ways of investigating the computability of problems, many of which are far more theoretical and abstract. However, all are equivalent to the Turing machine concept. We will not demonstrate the equivalence in this book. Instead, we refer you to Mendelson's *Introduction to Mathematical Logic* (Van Nostrand, 1964) or Bechman's *Mathematical Foundations of Programming* (Addison-Wesley, 1970).

RECURSIVE FUNCTIONS

One of the purposes of our examination of Turing machines was to investigate how we could construct a machine that would emulate the behavior of a function. Functions computable by a Turing machine are known as **Turing—computable functions**. Other approaches to computability use other vehicles to emulate functions. Register machines, for example, are constructs similar to Turing machines; they behave the way unbounded length registers of a computer might behave, and the functional behavior is described by the workings of the registers. Markov algorithms are more closely akin to the grammars we saw in the last chapter, and we examine these algorithms later in this chapter. Other computability structures are more esoteric and less like the structures with which you are already familiar. Conway's Game of Life, for example, is concerned with the generation of patterns in a plane. Another geometric approach is known as the tesselation of the plane. Tesselation involves using domino—like tiles in various arrangements to generate functions. Cellular automata are even more unlike the computer—like constructs we have seen; they were inspired by biological concepts. Other approaches abound, including Post canonical systems, Thue systems, and semi—Thue systems.

In this section, we examine another set of functions equivalent to the Turing—computable functions, namely, the recursive functions. Although the recursive functions are purely theoretical (that is, nonmechanical), they can be implemented fairly easily by recursive programming languages such as PL/I or Pascal. Their more theoretical nature presents quite a contrast to the Turing machines.

The set of recursive functions is generated by a rigorous and carefully defined formal system, much as are the Turing—computable functions above. In this case, however, the formal system is generated by defining a set of initial functions and applying rules that govern the generation of new functions. The axioms of this system are the initial functions, and the rules of generation are the rules of derivation of the formal system. The rules generate new theorems, which in this case are functions. Thus, in the discussion that follows, we will use the term "theorem" synonymously with "function" to remind you that we are indeed generating new results in a formal system.

The functions with which we will be dealing in our formal system are all defined over the same domain, namely, the set of positive integers. A subset S of I is said to be **recursively enumerable** if S is the empty set or if there is an algorithm F defined on I whose range is exactly S. (How does this definition differ from that of denumerable in Chapter 4?) For this reason, the functions in our formal system are known as the **recursively enumerable** or **recursive functions**. We can enumerate the elements of S by calculating $F(i)$ for every element i in I. (F need not be one—to—one.) Thus, the range of a recursive function is called a **recursively enumerable set**.

To define this formal system, we first define the set of initial functions from which we will generate the others. The initial functions are as follows:

1. The constant—zero function: This function is defined by $K_0(n) = 0$ for all positive integers n.
2. The successor function: For each positive integer n, this function defines its successor, $n + 1$: $S(n) = n + 1$.
3. The projection function: For each sequence of n integers, this set of n functions picks out the kth element, for k ranging from 1 through n: $P(n, k)(x_1, x_2, x_3, ..., x_n) = x_k$.

The rules for generating new theorems from these initial functions are described as follows:

1. Composition: We can substitute one theorem (i.e., function) into another. Thus, if H and G are theorems, then $F(n) = H[G(n)]$ is a theorem.
2. Recursion: If G and H are theorems, and we can define F as

$F(x_1, x_2, ..., x_m, 0) = G (x_1, x_2, ..., x_m)$

$F (x_1, x_2, ..., x_m, n + 1) = H [x_1, x_2, ..., x_m, n, F (x_1, x_2, ..., x_m, n)]$

then F is a theorem. This says that if F is definable in terms of the previous values of F, then F is a recursive theorem. In other words, F is a recursive theorem if $F (n)$ is a recurrence relation. We can express this as

$F (0) = k$

$F (n + 1) = H [n, F (n)]$ for n greater than 0 (The function F defined here is a 1=place function. However, the more general case is the $m + 1$=place function defined above.)

3. The μ−operator: Given an $m + 1$−place function $F(x_1, x_2, ..., x_m, y)$, the μ−operator operates by picking the smallest value of y so that the function value is zero. Thus, $\mu_y F(x_1, x_2, ..., x_m, y)$ is the smallest value of y such that $F(x_1, x_2, ..., x_m, y) = 0$. If there is no such smallest y, then the value of μ_y is undefined. For example, $\mu_y S (y)$ is the smallest y such that $y + 1$ is equal to zero. From the definition of S, it is easy to see that there is no such y. If F is a theorem, then $\mu_y F$ is a theorem. In defining $\mu_y F$, the value of y can be restricted to all those values of y beneath some upper bound. We write

$$\mu_{y \leq k} F$$

to denote the smallest y less than or equal to k such that F is equal to zero. If there is no such y, then the value of the function can be defined as zero, $k + 1$, or some other suitable value. This function is called the **bounded μ − operator**; it generates the set of **primitive recursive functions,** which form a proper subset of the recursive functions.

Just as it seemed unlikely that the set of Turing machines could generate the set of all programs, it seems equally unlikely that this set of initial functions and rules can generate the set of computable functions. Nevertheless, it is true. Let us look in detail at some examples of how new functions are generated from the initial set.

The constant−1 function is the function which is 1 for all values of n; that is, $K_1 (n) = 1$ for all n. This function can be generated as

$K_1 (n) = S[K_0 (n)] = 1$

As you can see, we applied the successor function to the constant−0 function to get the constant−1 function.

Let $K_{1,2}$ (m, n) represent the two−place constant−1 function. This function is to be equal to 1 for any pair of values (m, n). We can create this function by using a combination of the successor function, the constant−0 function and a projection function in the following way:

$$K_{1,2}\ (m, n) = S[K_0\ (P_{2,1}\ (m, n))] = S[K_0\ (m)] = 1$$

This sequence uses the projection function to pick out the first element of the pair; then, the constant−0 function transforms the element to a 0; finally, the successor function yields a 1.

The third example is sometimes known as the **sign** function. It generates a 0 if the input is zero and a 1 otherwise. The sign function, denoted by Sg, is derived by setting

$$Sg\ (0) = K_0\ (0)$$

and

$$Sg\ (n\ + 1) = K_{1,2}\ [\ n,\ Sg\ (n)] = 1$$

This is an example of a recursive definition of a function. To generate a value for $Sg(n)$ when n is nonzero, it is necessary to know the values for Sg for all values less than n.

The predecessor function is defined to be

$$Pr(n) = 0 \text{ if } n \text{ is zero}$$
$$= n-\ 1 \text{ if } n \text{ is larger than zero}$$

We can form this function by using the constant−0 function for the first case:

$$Pr\ (0) = K_0\ (0) = 0$$

and the projection function for the second:

$$Pr(n\ + 1) = P_{2,1}\ [\ n,\ Pr\ (n)] = n$$

Again, this is a recursive definition.

The function $S(n, j)$ is a successor function on the jth element of an n-tuple. It looks like

$$S(n, j)(x_1, x_2, ..., x_j, ..., x_n) = x_j + 1$$

and can be generated in our formal system as

$$S(n, j)(x_1, x_2, ..., x_j, ..., x_n) = S[(P(n, j)(x_1, ..., {}_n)]$$

The summation of two numbers is computed by the summation function. It is normally expressed as

$$E(x,y) = x + y$$

But in our system, it is written as

$$E(x, 0) = Pr\ (1, 1)(x) = x$$

and

$$E(x, n + 1) = S\ (3,3)[\ x, n, E(x, n)] = E(x, n) + 1$$

Now, let us look at the function which computes the product of two numbers. Instead of writing it as

$$\Pi\ (x, y) = x \times y$$

we instead write

$$\Pi\ (x, 0) = K_0\ (x) = 0$$

and

$$\Pi\ (x, n + 1) = E\ [3,(1,3)][\ x,n,\ \Pi\ (x,n)]$$

where $E\ [3, (1, 3)]$ is defined similarly to E and specifies that the first and third items of a 3-tuple are to be added. Using this definition, we have

$$\Pi\ (x, n+1) = E[P\ (3, 1)[\ x, n, \Pi\ (x, n)], P\ (3, 3)[\ x, n, \Pi\ (x, n)]]$$

As a final example, we present a function well known as the **Ackermann function**. It is defined recursively as

$$ACK(0, N) = N + 1$$
$$ACK(M + 1, 0) = ACK(M, 1)$$
$$ACK(M + 1, N + 1) = ACK[M, ACK(M + 1, N)]$$

Using a recursive language (such as Pascal or PL/I), a subroutine to calculate the values of this function would use less than 10 lines of code. On the other hand, this function is very difficult to implement in a nonrecursive language (such as FORTRAN — try it!). Although the Ackermann function is deceptively simple, it is not a primitive recursive function. In other words, it cannot be generated without the unbounded μ — operator. This function is thus used to show that the primitive recursive functions are a proper subset of the recursive functions.

It is interesting to note that the values of this function grow large very quickly. Even with a large memory capacity and a good deal of CPU time available, your computer may not be able to calculate some of the values of the Ackermann function without using some tricks (such as saving intermediate values in an array). The problem stems from the third line of the function definition. For instance, to calculate $ACK(6, 6)$, you know that

$$ACK(6, 6) = ACK[5, ACK(6, 5)]$$

Since the values of this function grow large quickly, the value for $ACK(6, 5)$ is quite large. This means that we have

$$ACK(6, 6) = ACK(5, \text{very large number})$$

You can see that filling in an array can grow to be a FORTRAN nightmare, and that using a recursive language is almost a necessity. On the other hand, even though a short and elegant Pascal program may guarantee the computability of a number such as $ACK(50, 50)$, it may take your computer an inordinate amount of time to actually produce the number.

MARKOV ALGORITHMS

We will investigate one last set of constructs in which computable problems can be formulated. This is the set of Markov algorithms. Like recursive functions, Markov algorithms are composed of a set of rules, but unlike the recursive functions, the set of rules is ordered. **Markov algorithms** are an ordered sequence of rules in which the rules are applied to strings to form other strings. The lowest—numbered rule is applied at each stage of the generation, and the rule is applied as far to the left as possible in the input string. An auxiliary alphabet is allowed, just as we had one for grammars in the previous chapter. The symbol $\wedge$ represents the null or empty string; it matches any location in any string. The symbol * is used to represent a stop. Let us look at some examples to see how the rules and symbols are used. In our example, small letters represent the regular alphabet, while capital letters are the letters in the auxiliary alphabet.

Suppose we are given a string with n occurrences of the letter a, and we want to form a string with $2\,n$ occurrences of a. The following rules we can use to perform this transformation.

1. $Aa \rightarrow aaA$
2. $A \rightarrow \wedge*$
3. $\wedge \rightarrow A$

To see how these rules work to generate new strings, consider a string of n a s: $aaaa \ldots aaa$. Rule 1 cannot be applied because there is no A present in this string; likewise for rule 2. Thus, we must apply rule 3. Applying rule 3 to the leftmost symbol, we have $Aaaa \ldots aaa$. Now, we keep applying rule 1 until we have doubled the number of as and reach the string $aaa \ldots aaaA$. At this point, we can no longer apply rule 1, so we must go on to apply rule 2. This erases the symbol $\wedge$ and stops.

Suppose we want to create a set of Markov algorithms to accept a string of as and bs such as $abaabbaaabba$ and separate them into first the as and then the bs: $aaaaaaabbbbb$. We can use a set of rules that look like this:

1. $Ba \rightarrow aB$
2. $Bb \rightarrow bB$
3. $B \rightarrow C$
4. $Aa \rightarrow aA$
5. $Da \rightarrow aD$
6. $Db \rightarrow bD$
7. $DC \rightarrow Cb$
8. $AC \rightarrow \wedge*$
9. $Ab \rightarrow AD$
10. $\wedge \rightarrow AB$

Given a sequence of as and bs, none of the rules applies until we get to rule 10. Applying rule 10 results in

$ABabaabbaaabba$

Next, we apply rules 1 and 2 to get

$AabaabbaaabbaB$

We then apply rule 3 to get

   AabaabbaaabbaC

If we apply rule 4, we end up with

   aAbaabbaaabbaC

Had there been no *b*s, we would have yielded a string of the form

   aaa ...aaaAC

Now we are in a position where only rule 9 applies. Applying rule 9 changes *Ab* into *AD*. Continued application of this rule results in the *D* 's moving to the right, passing the *a*s and the *b*s until the *D* reaches the *C*. At this point, rule 7 can be used, and this changes *DC* into *Cb*. If we continue in this fashion, we end up with

   aaaaaaaACbbbbb

whereupon we apply rule 8 and stop.

WHEN IS SOMETHING COMPUTABLE?

In the previous sections, you have seen three ways of describing problems: in terms of Turing machines, as recursive functions, and as Markov algorithms. All of these expressions of a problem are equivalent, and they can help us to decide whether a problem is computable. In general, a problem is computable if

   It can be described in terms of a Turing machine that halts.

or

   It can be defined completely as a recursive function.

or

   It can be described as a set of Markov algorithms that stop.

These structures are thus quite powerful. Although it is not an easy task to express a typical problem in one of these three contexts, it is usually simpler to examine a problem once it is in one of these frameworks. That is why computer scientists continue to investigate the properties and implications of Turing machines, recursive functions, and Markov algorithms.

EXERCISES

1. We say that a procedure is **effective** procedure if it is a process whose execution is specified totally and clearly. A problem P is **semidecidable** if there is an effective procedure F which, when applied to a question q of a problem P, produces the answer *yes* and halts if and only if q has the answer yes. F need not halt if q has answer *no*. Show that a problem P is recursively enumerable if and only if P is semidecidable.

2. a. Show that if a problem P is solved, P is solvable.
   b. Show that if P is solvable, P is recursively enumerable.

3. Show that if a problem P is finite, it must be solvable.

4. Design a Turing machine to multiply two numbers. Your Turing machine will start with two numbers, N_1 and N_2, separated by a blank. The result will be the number $N_3 = N_1 \times N_2$.

5. Design a Turing machine to determine the integer quotient of two numbers. The machine will start with two numbers, N_1 and N_2, separated by a blank. The result will be the greatest integer less than or equal to the quotient N_1 / N_2.

6. Design a Turing machine to determine the remainder after dividing one integer by another. The machine will start with two numbers, N_1 and N_2, separated by a blank. The result will be the remainder after calculating N_1 / N_2.

7. The difference function $D(x, y)$ computes the difference of two numbers, x and y, in the following manner:

$$D(x, y) = x - y \text{ for } x \text{ greater than or equal to } y$$
$$= 0 \text{ for } y \text{ greater than or equal to } x$$

How is the difference function expressed in recursive functions defined as in this chapter?

8. How are the following functions expressed in the formal system of recursive functions?

   a. $A(x, y) = |x - y|$

b. $Q(x, y)$ = the integer quotient of x/y (Hint: use the μ —operator.)

c. $R(x, y)$ = the remainder in x/y

d. $SQ(X)$ = the largest integer in the square root of x (**Hint:** $SQ(x)$ is the smallest integer N such that $N^2 \leq x$ and $(N + 1)^2 > x$.)

e. $FACT(n)$ = n ! (**Hint:** $FACT(n + 1) = FACT(n) \times S(n)$.)

9. Give a recursive function $PM(n)$ so that

$PM(n) = 1$ if n is prime

$PM(n) = 0$ if n is not a prime number

10. Produce a recursive function equivalent to the function $EXP(m, n)$ = m^n. Define 0^0 as 1.

11. Produce a Markov algorithm to calculate the quotient N_1 / N_2, where the numbers N_1 and N_2 are represented by the following string:

$$\longleftarrow N_1 \longrightarrow \quad \longleftarrow N_2 \longrightarrow$$
$$aaaaaaaaaaaaaaaabaaaaaaaaaaaaaaaaaaa$$

12. Devise a Markov algorithm to calculate the remainder after having divided N_1 by N_2, where N_1 and N_2 are represented as in the previous problem.

13. A palindrome is a sequence of characters which reads the same backward as forward. Produce a Markov algorithm to check for palindromes of *a*s and *b*s. If the input is a palindrome, the algorithm will stop. If the input is not a palindrome, the algorithm will erase the input and then stop.

INDEX

Absolute value function, 77
Absorption, 92
Ackermann function, 343
Acyclic graph, 244
Adjacency matrix, 259
Adjacent lines, 230
Adjacent terms, 116
Alphabet, 285
Ambiguous grammar, 315
Analog, 2
Ancestral of relation, 86
AND, 25
Antecedent, 34
Antisymmetry, 56, 153
Arc, 261
Argument, 38
Associativity, 73, 92, 159
Asymmetric relation, 56
Automorphism, 84
Auxiliary alphabet, 287
Axiomatic system, 21
Axioms, 21

Backing up flow, 274
Backus Naur form, 248
Bijection, 79
Binary addition, 105
Binary coded decimal (BCD), 126
Binary operation, 68
Binary relation, 55
Binomial coefficients, 211
Binomial theorem, 211
Bipartite graph, 243
Blocks of partition, 75
Boolean algebra, 91, 161
Boolean connective, 97
Boolean function, 96
Boolean operator, 97
Boolean variable, 96
Boundary condition, 220
Bounded μ — operator, 341
Bus route problem, 205, 229

$C(n, r)$, 208
Canonical product of sums, 108
Canonical sum of products, 108
Cardinality of set, 189
Cartesian product of sets, 53
Cellular automata, 339
Chromatic number of graph, 256

Circular arrangement, 204
Circular relation, 85
Closed form solution, 220
Closed walk, 236
Closure:
  of operation in set, 82
  of relation in set, 62
Coloring graph, 255
Combinational logic, 140
Combinations of propositions, 24
Common elements, 119
Communications network problem, 6, 263
Commutativity, 70, 73, 92, 159
Compiler correctness problem, 7
Complement, 20, 69, 92, 162
Complementive lattice, 162
Complete graph, 239
  bipartite graph, 243
Component of graph, 238
Composition, 340
  of two functions, 87
  of two relations, 85
Compound propositions, 28
Computability theory, 3
Computable problem, 346
Conclusion, 34
Conditional statement, 34
Congruence modulo n, 59
Conjunction, 25
Conjunctive normal form, 108, 201
Connected graph, 237
Consequent, 34
Constant-zero function, 340
Context, 23
Context-free grammar, 289
Context-free language, 289, 310
Context-sensitive grammar, 290
Context-sensitive language, 289, 313
Continuous, 2
Continuum hypothesis, 196
Contrapositive of implication, 35, 36
Converse:
  of implication, 35
  of relation, 85
Converting NFSA to DFSA, 304
Conway's Game of Life, 339
Correct propositions, 33
Correspondence, 53
Countable set, 191